ARCHITECTURAL
PRECAST CONCRETE
DRAFTING HANDBOOK

ARCHITECTURAL PRECAST CONCRETE DRAFTING HANDBOOK

Prepared by
PCI Committee on Architectural Precast

Concrete Shop Drawings
Robert W. Johnson, Chairman

John R. Chesterfield
Jerry Cooksey
Stephen D. Disch
R. J. Jarvis

Raymond Neuhaus
F. Parakovits
Cal Saari
Burton Stone

Prentice-Hall, Inc., *Englewood Cliffs, New Jersey*

Library of Congress Cataloging in Publication Data

Prestressed Concrete Institute. Committee on
   Architectural Precast Concrete Shop Drawings.
   PCI architectural precast concrete drafting
handbook

   1. Architectural drawing—Handbooks, manuals, etc.
2. Precast concrete construction—Handbooks, manuals,
etc. I. Title.
NA2705.P73 1975   720'.28   74-26663
ISBN 0-13-044602-5

© 1975 by
Prentice-Hall, Inc.
Englewood Cliffs, N.J. 07632

10 9 8 7 6 5 4 3 2 1

Printed in the United States of America

PRENTICE-HALL INTERNATIONAL, Inc., London
PRENTICE-HALL OF AUSTRALIA, Pty Ltd., Sidney
PRENTICE-HALL OF CANADA, Ltd., Toronto
PRENTICE-HALL OF INDIA PRIVATE LIMITED, New Delhi
PRENTICE-HALL OF JAPAN, Inc., Tokyo

CONTENTS

II DRAWING PREPARATION   15

III THE SUBMITTAL PROCESS   83

IV THE DRAFTING ROOM   91

APPENDICES   *109*

FOREWORD

The Prestressed Concrete Institute, a non-profit corporation, was founded in 1954 for the purpose of advancing the design, manufacture, and use of prestressed and precast concrete. The Institute represents the prestressed concrete industry, and a large segment of the architectural precast concrete industry, in the United States and Canada.

The many technical, research, and development programs of the Institute are financially supported, primarily, by about 200 Company Members, representing some 350 plants throughout the United States, Canada, and many foreign countries. These firms are engaged in the manufacture of precast and prestressed concrete products and prestressing materials for the construction industry.

Important financial and technical support is also received from Associate Member Companies, supplying materials and services related to the industry, and from individual Professional Members engaged in the practice of engineering, architecture, or related professions. Affiliate and Student Members are other important categories of individual PCI membership.

As the spokesman for the prestressed and precast concrete industry, PCI continually disseminates information on the latest concepts, techniques, and design data pertinent to the industry and to the architectural and engineering professions through regional and national programs and technical publications. These programs and publications are aimed at advancing the state of the art for the entire industry.

Engineers, architects, draftsmen, contractors, and owners interested in prestressed and precast concrete design and construction, should contact PCI headquarters, 20 North Wacker Drive, Chicago, Illinois 60606 for information on all aspects of such construction as well as for information on PCI membership.

PREFACE

The acceptance of the architectural use of precast concrete by architects, engineers, and builders began to flourish at the end of World War II. Since that time, we have witnessed its acceptance by the architectural profession on a world-wide scale. The industry's phenomenal growth has necessitated the development of guidelines for the preparation of architectural precast concrete drawings which would lead to eventual standardization of these drawings nationally in terms of format, information to be included, assignment of responsibilities, symbols for hardware and other requirements, and procedures for preferred stages of drawing submittal.

The term precast concrete as it applies to building construction is divided into two major categories: architectural and structural. This handbook concerns itself only with conventionally reinforced architectural precast applications. Prestressed panels are not discussed.

Herein, the term architectural precast concrete

encompasses any precast concrete unit employed as an element of architectural design, whether structural or decorative, and made in custom-designed shapes, sizes, and finishes for the individual project.

The primary function of precast erection and production drawings* is the translation of contract documents into usable information for accurate and efficient manufacture, handling, and erection of the precast concrete units. In addition, the drawings provide the architect with a means of checking interfacing with adjacent materials. Good production drawings reduce plant costs and speed production by providing effective communications between the engineering-drafting and the production-erection departments of a precast plant. Additionally, the erection drawings provide the precaster with his only practical means of communicating with the builder and architect.

On these pages—for the first time—there appears a comprehensive text on the subject of Architectural Precast Concrete Drafting. Compiled and edited by the Prestressed Concrete Institute, this handbook presents a single authoritative reference for the architectural precast concrete draftsman. To supplement the uniform drawing standards, general information on precast concrete drafting procedures and techniques has been included. This information augments basic drafting techniques.

PCI has long recognized the need for a handbook which provides the guidelines and recommendations pertinent to drafting practices in the Architectural Precast Concrete Industry. The preparation of the text of this handbook was by a small committee made up of nine representatives from both PCI Member Companies of varying sizes and consulting firms (PCI Professional Members). Each individual has many years of experience in architectural precast drafting.

PCI *Architectural Precast Concrete Drafting Handbook* is intended to serve a dual function: (1) as the industry standard for use by any individual involved in the preparation and/or use of precast drawings, and (2) as a textbook for use in training draftsmen in high schools, vocational schools, and on-the-job

---

*See page 20 for explanation for types of precast concrete drawings.

training programs conducted by architectural precast concrete producers or drafting firms.

All of the information presented shows current recommended practices in the industry. While every precaution has been taken to assure that all data and information presented is as accurate as possible, PCI does not assume responsibility for errors or oversights in the information published herein, in the use of such information, or in preparation of precast drawings incorporating such information. *Under no circumstances should any of the information herein be used to make design decisions.*

Each procedure described can be effectively utilized to generate architectural precast drawings successfully. However, it would be an expensive duplication of effort to utilize all of the drawings described in the handbook for each and every project, regardless of size. The reader must determine personally the drawing requirements for a given project.

For successful detailing, it is desirable for the draftsman to have had experience with the various processes of production and erection.

A short period of work in each of the appropriate departments greatly helps in achieving an understanding of the practical aspects involved, and assists the draftsman in appreciating the part played by every employee in the successful manufacture and erection of the architectural precast concrete. The qualified draftsman must also be acquainted with many details and construction methods used in conjunction with other types of material which may be installed (interfaced) with architectural precast concrete.

The ability to be a good draftsman is not an innate talent; it can be acquired by hard work, practice, diligence to detail, and a self-imposed insistence on neatness. Once acquired, drafting proficiency will be maintained only by constant effort.

PCI appreciates the considerable time and effort expended by individuals from the following firms (in alphabetical order) to develop this handbook: Ace Concrete Company, American Precast Concrete, Inc., The Babcock Company, T. J. Escedi & Associates, Ltd., Francon Division of Canfarge Ltd., Olympian Stone Company, Inc., Raths, Raths & Johnson, Inc., Southern Cast Stone Co., Inc., Spancrete Midwest Co., and Carl Walker & Associates, Inc.

DRAFTING TECHNIQUES

AND PROCEDURES

Application of proper drafting techniques is important to the successful completion of any drawing. A drawing is more than a graphic delineation of an object or structure; it is a definite order to workmen to perform certain operations in a specified manner. This is especially true for architectural precast concrete drawings.

Section I

1

Expression in drafting concerns itself mainly with the manner and technique employed by the draftsman in conveying correct information on drawings simply and clearly. Each drawing should be completed in a step by step fashion, beginning with object delineation, proceeding through dimensioning, and culminating with lettering applications. Drawings should not contain unnecessary lines, marks, symbols, or dimensions. *However, they must contain an adequate set of notes, and all other essential information, in a form that can be quickly and correctly interpreted.* It helps if, as he works, the draftsman puts himself in the place of the production personnel or erection crews who must search the drawings for each bit of information. The following points on draftsmanship will be helpful in communicating ideas on paper with a minimum of misinterpretation.

Views

Although pictorial drawing—isometric or oblique views—has some application in developing and communicating ideas, it does not lend itself readily to making precast concrete drawings. Consequently, a multiview system known as orthographic projection is used for drafting throughout the industry. The basis of this method is to show the characteristics of an object by using as many dimensioned views as necessary to describe it fully. The views show the shape of the object as observed from several directions, and are related to each other by location on the drawing and by their dimensioning.

In precast concrete drafting, views can be separated into three categories. (See figure 1-1.)

PRIMARY—plan, elevation, section
SECONDARY—top, side, end, back, front, section
AUXILIARY—detail, blow-up, exploded, isometric

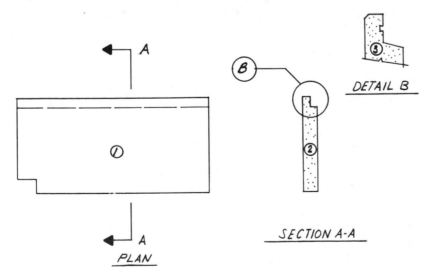

Figure 1-1

A *Primary* view, as the name implies, is the main view from which all other views are projected. *Secondary* views are two-dimensional orthographic projections taken from the primary view. *Auxiliary* views are those added for clarity.

Lines   Clarity and ease of reading are increased if the same type of line is used throughout for the same purpose. The weight or width of each line should be varied to accentuate important features. Dimension/extension lines, while being the thinnest lines on the drawings, must be dense enough to reproduce clearly when multi-generation sepias are made. Standardization of line work for all drawings will also speed up preparation and use of the drawing.

All line work employed in precast concrete drafting falls into one of these eight categories:

1. Object	5. Leader
2. Hidden	6. Center
3. Extension	7. Break
4. Dimension (primary, secondary)	8. Symbols

Figure 1-2 illustrates the appearance of such lines as they relate to one another on a drawing.

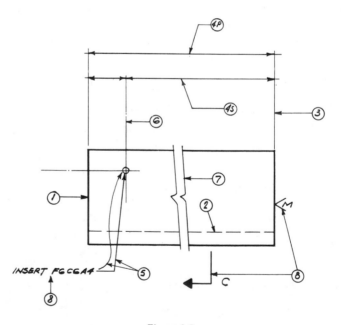

Figure 1-2

Lettering   All notes, letters, dimensions, and arrowheads used on drawings should be made freehand, using a style that is legible, uniform, and capable of rapid execution. Individual preferences should indicate whether vertical or slanted lettering be employed. The slanted style is used throughout this handbook. The height and boldness of lettering and numerals should be in proportion to the importance of the note of dimension. For titles 3/16 inch to 1/4 inch is adequate while 1/8 inch should be used for notes. *Under no circumstances, should lettering guides be employed as they tend to reduce productivity.* See figure 1-3 for examples of vertical and slanted lettering as well as a guide to lettering sizes for normal applications.

ABCDEFGHIJKLMNOPQRSTUVW XYZ 1234567890 $\frac{13}{64}$ = VERTICAL

ABCDEFGHIJKLMNOPQRSTUVW XYZ 1234567890 $\frac{13}{64}$ = SLANTED

TITLES   DIMENSIONS, NOTES

Figure 1-3

Two types of dimensioning are employed within the precast concrete industry. They are: (1) point-to-point, which is the method used in this handbook, and (2) continuous dimensioning, which is recommended only as a supplement to point-to-point dimensioning. Point-to-point dimensioning relates a point to the base (zero) line. It may have a number of "zero" bases. Although continuous dimensioning minimizes the possibility of cumulative error in the shop by relating all dimensions to a single "zero" base, a major drawback is that it requires subtraction to find the distance between any two points. The possibility of errors in drafting is therefore increased. See figure 1-4 for examples.

The following are recommended dimensioning practices and cover most conditions normally encountered:

1. Horizontal and sloping dimensions always read from left to right, and vertical dimensions from bottom to top. It is imperative that the drawing be readable from the bottom and the right side to enable plans to be read without turning them. (See figure 1-5.) All notes should read horizontally, when possible.

2. Place all dimensions outside of views. Only if added clearness and simplicity results should they be placed within a view. Do not place dimensions on cut surfaces of a section. If it

Dimensioning Practices

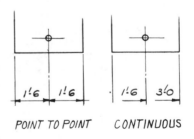

POINT TO POINT   CONTINUOUS

Figure 1-4

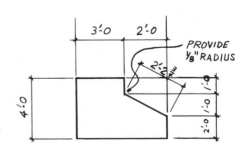

Figure 1-5

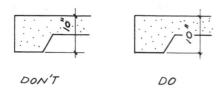

DON'T DO

Figure 1-6

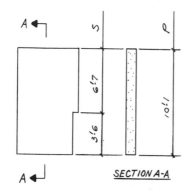

Figure 1-7

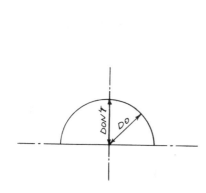

Figure 1-10

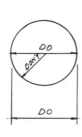

Figure 1-12

becomes necessary to place a dimension on a view, all section lining should be omitted around the numbers as shown in figure 1-6.

3. *Primary dimensions should be placed outside of the views on the extreme dimension line.* When possible secondary dimensions should be placed between views, and always below or nearer the view than primary dimensions. (See figure 1-7.)

4. Parallel dimension lines should be kept at equal distances apart. (See figure 1-8.)

5. Figures should be placed midway between dimension points if possible (unless a centerline interferes or several parallel dimensions are staggered). Offset them slightly to avoid crossing extension lines. (See figure 1-8.)

6. Never place letters or numbers directly on linework nor allow them to touch lines on the drawing.

7. Dimensions to circular parts are always given from center to center. (See figure 1-9.)

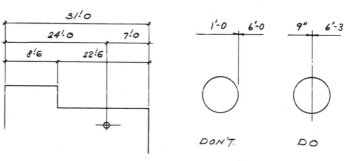

Figure 1-8

Figure 1-9

8. Centerlines are never used as dimension lines. (See figure 1-10.)

9. Object lines must never be used as dimension lines. (See figure 1-11.)

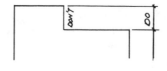

Figure 1-11

10. The diameter of a circular opening, not the radius, should be indicated on the drawing. This figure should be followed by the abbreviation or symbol for diameter. (See figure 1-12.)

11. When specifying the radius of an arc, follow the dimension with the abbreviation for radius. Radial dimension lines do not have arrowheads at their center. (See figure 1-13.)

Figure 1-13

12. When locating the center of an arc lying outside the limits of the drawing, the radial dimension line is offset. (Refer to figure 1-14.)

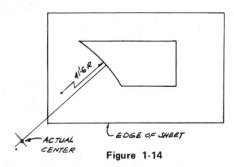

Figure 1-14

13. Avoid placing dimension lines in a direction included in the shaded area of figure 1-15. If this is unavoidable, they should read as shown.

14. When indicating a number of dimensions in a row, they could be placed in line or staggered. *To eliminate confusion it is suggested that staggered dimensioning be discouraged unless added clarity is achieved.* (See figure 1-16.)

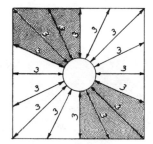

Figure 1-15

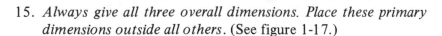

IN LINE   STAGGERED

Figure 1-16

15. *Always give all three overall dimensions. Place these primary dimensions outside all others.* (See figure 1-17.)

16. *Never require the reader to add or subtract dimensions or to scale a drawing.* (See figure 1-18.)

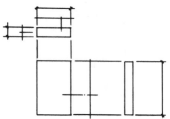

Figure 1-17

17. When placing dimensions in crowded areas, use an enlarged auxiliary view or, if possible, one of the procedures shown in figure 1-19.

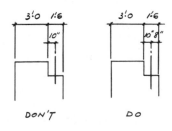

DON'T   DO

Figure 1-18

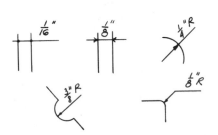

Figure 1-19

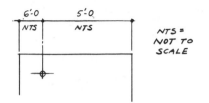

Figure 1-20

18. When dimensions are changed and are no longer in proportion, they should be noted as shown in figure 1-20.

19. Angle degree designations should be placed horizontally for small scale angles. Angles of larger scale may be dimensioned parallel with the arc. (See figure 1-21.)

20. Curved lines should be dimensioned either by radii or by offsets as shown in figure 1-22.

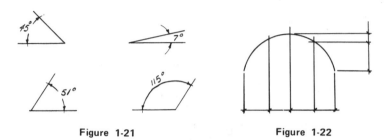

Figure 1-21   Figure 1-22

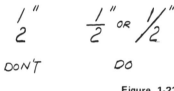

Figure 1-23

21. Always use a line between the numerator and denominator when expressing a fraction. (See figure 1-23.) Horizontal bars should be used unless restricted space demands the reduced fraction height made possible by sloping the bars. The bar of the latter style fraction should be so inclined and of sufficient length as not to be confused with the figure "1" in such fractions as 1/8.

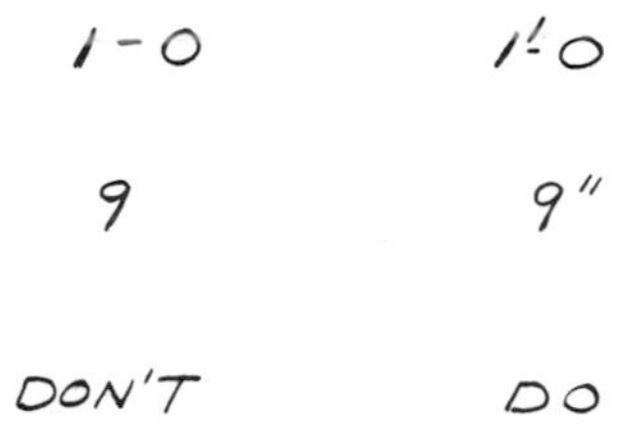

22. Always indicate the foot symbol when listing a dimension in feet. Inch symbols should be used only when no foot dimension exists. See figure 1-24 for recommended applications of foot and inch symbols.

23. Make ample sized decimal points.

24. Always dimension from a point easily accessible to shop personnel during the casting operation. (See figure 1-25.)

Figure 1-24

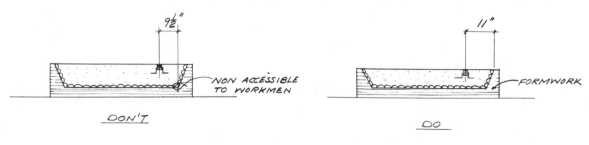

Figure 1-25

Freehand sketching is the accepted medium of communication between the draftsman and other individuals during drawing preparation. Proficiency in this area is normally acquired through years of instrument drafting. To effectively communicate intended messages, sketches should be carefully executed using clean, straight lines. Normal drafting techniques such as proper line weight, lettering, and dimensioning practices should be employed. Freehand drawings should be proportioned as close to scale as possible without the use of a scale. Gridded sketch pads are available to aid in freehand sketching (See figure 1-26.) During the printing process this grid does not reproduce. Give each sketch a sheet number and file it for future reference. Use SK for the sheet number prefix.

Sketching

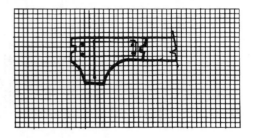

Figure 1-26

Proper use of reproducible and nonreproducible prints can save the precast draftsman many hours of work. In addition to effecting a short cut, this procedure can also be beneficial in minimizing mistakes caused through re-execution of repetitive work.

Drafting and Reproduction Timesavers

Nonreproducible Prints. When portions of completed drawings must be duplicated in whole or in part on a new drawing, it is common procedure to print the completed drawing and trace over the desired area. (See figure 1-27.) This eliminates the necessity of redrafting to scale.

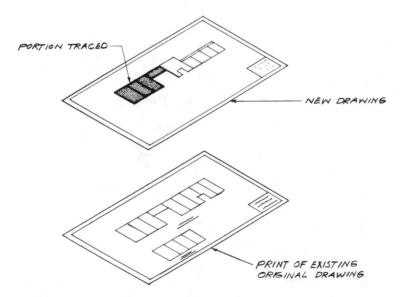

Figure 1-27

Reproducible Prints (Sepia). When several drawings are required and the differences are relatively minor, an original drawing should be prepared reflecting only line work and lettering which is common to all required drawings. Sepias (transparent prints) are then made for each drawing and each is then filled out with the required variations. (See figure 1-28.) This sepia technique will eliminate the need for extensive eradication or erasing. *It is highly recommended that the original be carefully checked prior to making the sepia prints.* This is to ensure that all repeated

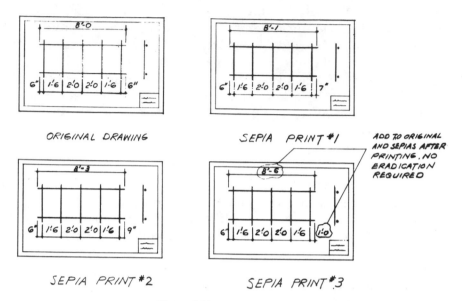

Figure 1-28

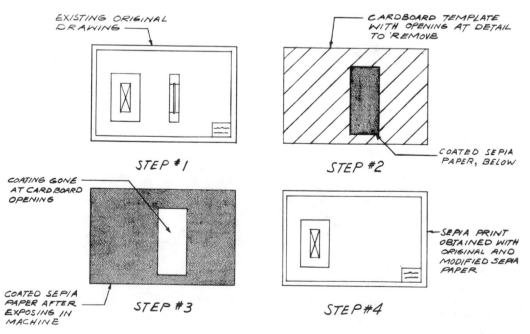

Figure 1-29

information is correct. Another application of sepia prints is to block out certain undesired portions of an existing drawing. (See figure 1-29.) Still another application is when typical details and sections are prepared on transparent paper, arranged on a blank sheet, copied via sepia print, and re-used for other drawings. (See figure 1-30.) Sepia prints may also be employed to replace damaged original tracings or to provide reproducible copies for approval. Maximum reprint quality is realized when reverse sepia prints are made. (See figure 1-31.) This also permits addition of details and dimensions to the front working surface while eradication or removal of undesired items takes place on the opposite side. (See figure 1-32.) This is important, as application of eradicator directly to pencil lines can result in blurring of these lines.

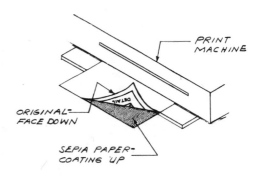

Figure 1-31

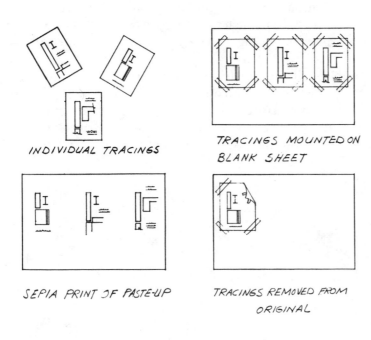

INDIVIDUAL TRACINGS

TRACINGS MOUNTED ON BLANK SHEET

SEPIA PRINT OF PASTE-UP

TRACINGS REMOVED FROM ORIGINAL

Figure 1-30

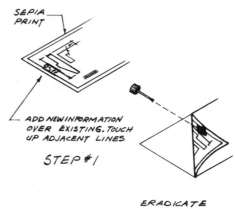

SEPIA PRINT

ADD NEW INFORMATION OVER EXISTING. TOUCH UP ADJACENT LINES

STEP #1

ERADICATE BACK OF PRINT

STEP #2

Figure 1-32

Eradication is often accomplished by placing an acidic chemical solution on the sepia surface. Common applicators include cotton swabs, brushes, and droppers. As soon as the image disappears, excess fluid is removed by use of an ink blotter. Some sepia papers are now available which can be altered with an electric eraser.

NOTE: All changes should be made to the front surface of the sepia, including redrawing and relettering of areas which may be affected by the eradicator. After all changes are made the eradicator is then applied to the back surface. This procedure is recommended since oftentimes eradication prior to changes results in failure to replace required information.

Stick-ons. When certain details are constantly repeated, it is recommended that they be reprinted on transparent, adhesive-backed film. These master templates can then be applied to any drawing, saving drafting time. Extensive notation should be typewritten on transparent adhesive film (available in rolls or sheet) and placed on the drawing. In addition to saving drafting time, this procedure has a tendency to yield neater and more legible notation, and to cut down on errors and omissions sometimes incurred when repetitive work is done. The frosted applique does leave a slight shadow on diazo reproductions.

Rubber Stamps. The use of rubber stamps can also be a valuable timesaving technique. In addition to numerous readily available stamps, custom-made stamps provide an endless variety of drafting aids. Some applications of rubber stamps are as follows: notation, details, drawing status indicators, and symbols. (See figure 1-33.)

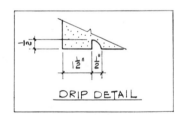

Figure 1-33

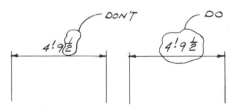

Figure 1-34

HINTS

a. When changing a dimension on a sepia, remove the entire number, not just the portion to be changed. This will help to eliminate errors. (See figure 1-34.)

b. When heavy lines are required a workable fixitive is recommended to prevent smearing. (See figure 1-35.)

c. Erase both sides of the sheet when line work and/or lettering is removed. This removes indentation and eliminates any trace of dirt which might have been fixed on the back of the sheet during the drafting process.

d. Avoid artistry by eliminating any unnecessary elaboration that adds nothing to the message.

e. Never draw any view or detail which is not required.

f. If something is easily described via notation, do not draw it.

g. Use dotted lines only for clarification.

h. Take full advantage of standard symbols.

i. Hold graphic notation to a minimum.

j. If freehand drawing is adequate, do not redraw with instruments.

k. Prior to printing, brush both sides of the original drawing to remove dirt. In addition to improving print quality this also helps to keep the print machine cylinder clean.

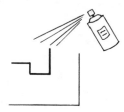

Figure 1-35

DRAWING PREPARATION

Preparation of clear, accurate, and complete precast drawings will save time and expense during the engineering, fabrication, and erection stages of a project. Since few, if any, precast jobs are even remotely similar, standardized methods for the preparation of precast drawings are a necessity.

Section II

The effectiveness of a drawing is measured in terms of how well it communicates its intentions to those who must rely on its message. To the user, an erection, or production drawing, is a detailed list of instructions in words as well as pictures for his guidance. With this thought in mind, the draftsman can materially improve the presentation by observing the following: (1) issue orders as commands—"bend bar" rather than "this bar is to be bent;" (2) emphasize the specific items delineated, i.e., in detailing reinforcing materials show lightly only enough of the surrounding precast unit to convey adjacent conditions, and show the reinforcing bars, welded wire fabric, and bar supports, (if they are included) heavily, clearly, simply, and completely; (3) make all notes on the drawings brief, clear, and explicit, leaving no chance for misunderstanding; (4) make all drawings large enough and letter them clearly enough so that after repeated handling and folding in the shop and/or field they will still be legible to the user. Drawings that will be reduced photographically require broader lines and coarser lettering. *Clarity is the most important consideration*. If a section or a plan becomes too crowded to detail a certain feature, select a larger drawing scale, or draw an auxiliary view of that particular area.

This chapter is intended to aid in the preparation of each type of precast drawing.

Pre-drawing Considerations

Prior to the start of any precast concrete project a thorough review and restudy of contract documents is necessary to determine all the factors that can influence decisions regarding the precast concrete. Contract documents usually encompass the architectural and structural plans, the project specifications, and the precast firm's proposal defining the conditions of the contract. It is most important that the contract proposal be carefully reviewed. Many times items are deleted or additional considerations are involved which can affect the Precaster's costs.

The contract documents should be checked carefully to discover any special conditions that may exist. For example, special conditions may include the hardware cast into the panels, the hardware used to attach the panels, the limitations posed by the project relative to the overall production (e.g., certain items required prior to others), or the specified delivery dates.

As soon as practical after award of the contract, and in accordance with the established contract schedule, the precast manufacturer should obtain all applicable contract documents, including structural, mechanical, electrical, and site drawings if necessary. Shop drawings from other related trades may also be needed. These must be coordinated in advance of precasting by the General Contractor to assure interfacing with such items as duct work, pipes, grills, openings, equipment foundations, and hangers. Also considered are provisions for concentrated loads and reactions. See Appendix G for a description of various types of loads and their application.

Before drawings are started, engineering personnel should meet with the production and erection staff to discuss overall job procedures as well as possible modifications or changes. At this time it may be advisable to meet with the Contractor and/or Architect to discuss and agree upon any desired changes. It is mandatory that a record of information and decisions made in the pre-drawing stages be available for distribution and also be retained in job file. (For Check List on drawing preparation see pages 76 to 82.)

Building Codes. It is essential that governing building codes be understood prior to the initiation of a project. It is also important to determine, in advance, what material or design requirements must be met.

Addenda. Contract drawings and specifications must be updated per the latest addenda.

Specifications. The specifications supplement working drawings to adequately define the end results expected from the architectural precast concrete, as well as all other building components. The precast section and general conditions must be thoroughly reviewed so that project specifications are fully understood. A cursory review of the other specification sections is also recommended.

Information contained in the specifications which must be noted includes:

Drawing Size	Material Requirements
Submittal Procedure	Precast Finish
Items specifically excluded	Waterproofing
Joint Sizes	Cast-in Items

Lead Time. It is advisable to determine—with the General Contractor—the amount of time available for completion of the precast drawings. This information will dictate the submittal sequence best suited to all parties involved. (See Section III, page 85.)

The importance of sufficient lead time is often overlooked. If recognized by the Architect, it affords cost savings in the precast portion and important savings in construction time for the overall project, as well as a better quality end product.

The speed with which the Contractor assembles the structure may indicate an early need for anchor plans or completion of certain areas. If the job is a multi-building complex, one building may be needed before others; therefore priorities must be set. *These and similar factors greatly affect the amount of lead time which, in turn, is directly related to production cost.*

Structural Frame. Prior to the laying of any lines, structural drawings must be fully reviewed to ascertain that all precast members have support. Should support conditions appear questionable or not be clearly understood, this fact must be brought to the attention of the Engineer of Record.

Production. The following are some of the production questions which must be answered:

1. Do precast units have adequate draft on exposed surfaces? (See page 55.)
2. Must handling inserts be recessed for patching due to exposure conditions?
3. Do forming problems exist which could be remedied without altering appearance of the panels?
4. Do special conditions exist which may require the attention of the Precast Engineer?
5. Are panels too large for ease of handling? (See Appendix O, page 182.)
6. Are cross sections of adequate size to receive reinforcement?
7. What type of finish is required for each surface?
8. What is the availability of specified materials and can readily available alternates be substituted for items difficult to obtain?

Shipping. The following are some of the important shipping items which should be taken into consideration:

1. Would a decrease in unit weight permit transporting two units per truck rather than a single unit?
2. Will the load be in excess of weight restrictions?
3. Do precast panel widths exceed legal limits?
4. Will loaded truck height exceed legal limits? (Low clearances enroute and at the job site should also be checked.)
5. Must special trailers be used? In some instances, the equipment used, and its configuration, influences the position of handling inserts or loops. The method of panel loading may dictate the length of handling loops. The type of trailers used in transit often dictates where panels are to be blocked for shipping. (See Appendix R, page 188.) The decision to ship a panel flat or vertically often is dictated entirely by the structural behavior of the panel when in these positions.

Erection. The following are some of the points which must be considered in the drawing stages to avoid complications at the job site.

1. Will cranes have adequate access to all parts of the building?
2. Must precast units be erected in a specific sequence?
3. Can units be combined to reduce erection time?
4. Can provisions for initial and long-term movements be built in throughout the building? (See Appendix E, page 155.)

Required changes or delays that are foreseen should be discussed with the General Contractor, so that future misunderstandings may be avoided.

Types of Drawings   Architectural precast drawings may be separated into two categories: Erection Drawings and Production Drawings.

Erection Drawings are prepared from the contract documents. They are submitted to the General Contractor, Architect, and Engineer of Record for approval, and must, therefore, clearly define the intent of the manufacturer. Following approval, these drawings are utilized in the preparation of shop drawings,* and again in the installation of precast units. For the Precaster's purposes, they actually replace the contract documents.

Types of Erection Drawings (Fig. 2-1a)

Cover Sheet	(see page 24)	Sections	(see page 37)
Key Plan	(see page 28)	Handling Detail	(see page 44)
Schedule	(see page 28)	Anchor Plan	(see page 46)
Elevation	(see page 29)	Hardware Detail (CIP & Loose)	(see page 50)
Erection Plan	(see page 34)		

Production Drawings are prepared from the erection drawings and the precast plant engineer's calculations. *Under no circumstances should information be obtained from contract documents for use in production drawing preparation.* The contract document information has not been approved for precast concrete production and may be incorrect or revised. Except for shape drawings, the production drawings are not submitted for approval, other than in special cases where the Architect, Engineer of Record, or the General Contractor agrees to assume responsibility. (See Section III and Appendix B.)

Production drawings serve as instructions to plant personnel for ordering, manufacturing, storage, and shipment of the precast product.

*Frequently, the term "shop drawings" is used in the precast concrete industry to refer to the production drawings or to both the erection and production drawings.

Types of Production Drawings (Fig. 2-1b)

Shape Drawing	(see page 53)	Bar List	(see page 68)
Panel Ticket	(see page 58)	Hardware Detail (Production)	(see page 50)
Reinforcing Ticket	(see page 64)	Material Lists (Production & Erection)	(see pages 71-72)

The distribution of the completed architectural precast drawings is shown in figure 2-2, and the specific relationships of the personnel involved in a project relative to their use of the drawings is shown in Table 2-1.

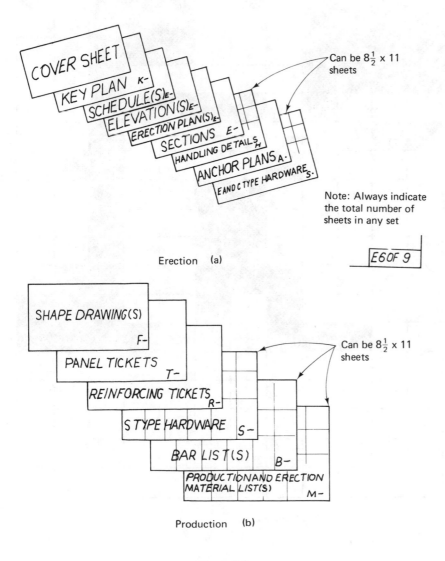

Erection   (a)

Production   (b)

Figure 2-1

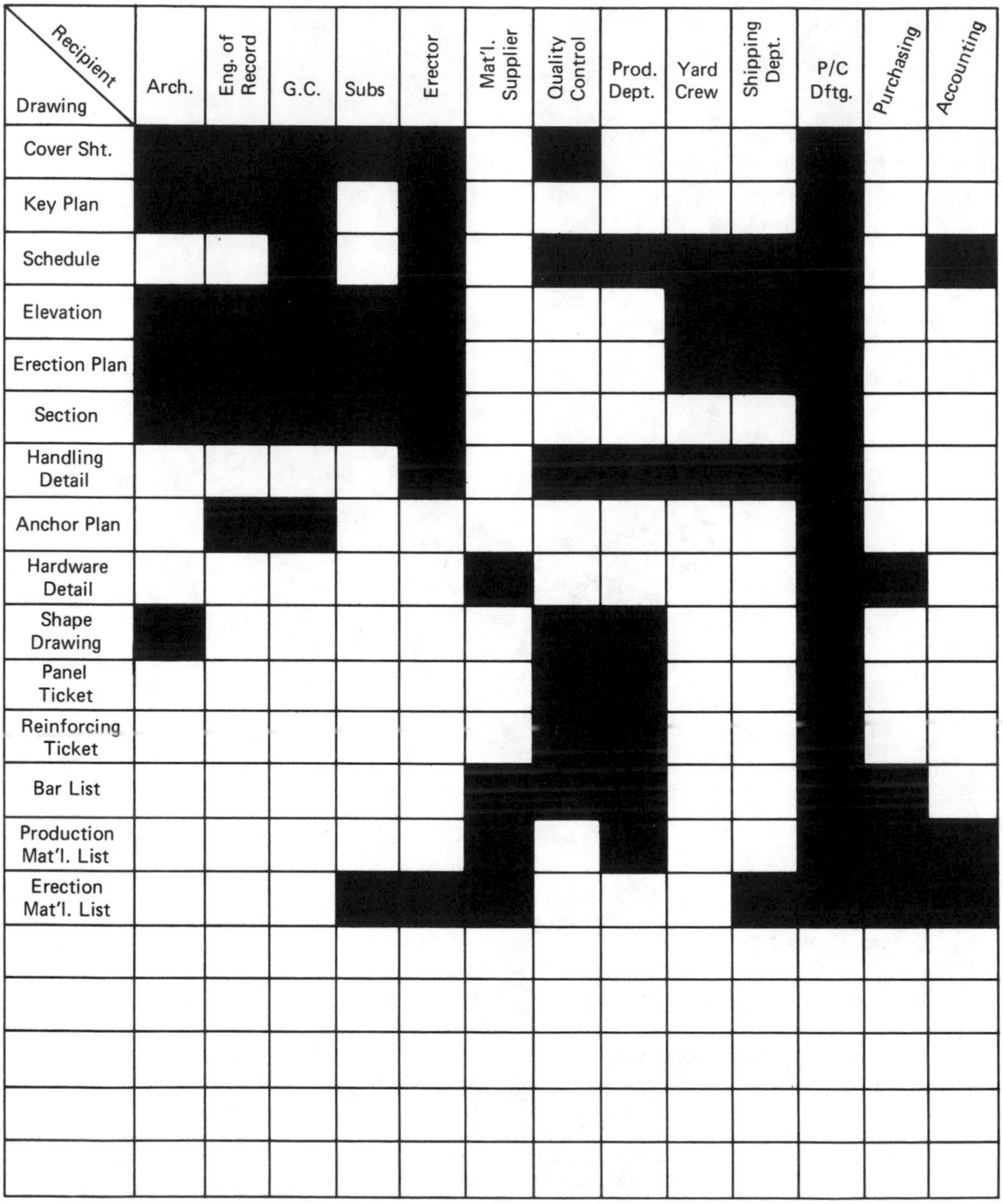

Drawing \ Recipient	Arch.	Eng. of Record	G.C.	Subs	Erector	Mat'l. Supplier	Quality Control	Prod. Dept.	Yard Crew	Shipping Dept.	P/C Dftg.	Purchasing	Accounting
Cover Sht.	■	■	■	■			■				■		
Key Plan	■	■	■		■						■		
Schedule			■		■		■	■	■	■	■		■
Elevation	■	■	■	■					■	■	■		
Erection Plan	■	■	■	■							■		
Section	■	■	■	■							■		
Handling Detail					■		■	■	■	■	■		
Anchor Plan		■	■					■			■		
Hardware Detail						■					■	■	
Shape Drawing	■						■	■			■		
Panel Ticket							■	■			■		
Reinforcing Ticket							■	■			■		
Bar List					■						■	■	
Production Mat'l. List				■	■			■			■		
Erection Mat'l. List				■	■		■			■	■	■	

Figure 2-2

Table 2-1.

	Function Relative to Project	Use of Precast Drawings
Architect	Initial conception and final design of building-through-planning and discussion with owner; writes specifications; coordinates bidding; works for owner.	Checks for conformance by precaster to his original intent.
Engineer of Record	Structural integrity of building; prepares design based on Architect's drawings; usually works for Architect.	Checks for conformace by Precaster to his original intent. Verifies that recommended changes do not adversely affect the structural integrity of the building.
General Contractor	Supervision of all construction in field; works for owner. Keeps Precaster supplied with up to date contract drawings and information on field alterations.	Helps in construction of building by coordinating precast with various trades. Ensures conformance to structure as built. Obtains cast-in-place anchor requirements.
Sub-contractor (electrical, plumbing, heating, ventilating, etc.)	Furnishes General Contractor with drawings and material from which the building is to be built; works for General Contractor. (Precasters generally fall into this category.)	Coordinates his trade's requirements with Precaster
Erector (precast)	Places precast units into their predetermined positions in the structure; may be general contractor; works for General Contractor or Precaster	Obtains loose hardware; determines crane requirements; determines setting sequence with the General Contractor and Precaster. Schedules his work.
Supplier (material)	Furnishes Precaster with materials required to manufacture precast panels; works for Precaster.	Finds details and quantity of material to fabricate and supply.

Precast Plant Personnel	Function Relative to Project	Use of Precast Drawings
Purchasing	Buys and coordinates delivery of material.	Obtains correct quantities of various materials.
Accounting	Keeps track of costs incurred.	Determines cost of individual units.
Quality Control	Ensures that finished products are in keeping with the tolerances and requirements set out in the erection and production drawings.	Checks forms, casting, set-up, reinforcing and completed units for conformance to production drawings.
Production	Manufactures units per production drawings.	Collects material and assembles into proper arrangement; notes handling procedures.
Planning and Scheduling	Schedules production.	Plans shop functions.
Yard	Finishes and stores units.	Obtains finish information; determines stacking sequence.
Shipping	Loads and transports to job site.	Arranges truck loads as dictated by erection sequence, weight, and local shipping laws.

Each type of erection and production drawing will be discussed in detail covering the following information where applicable: (1) planning, (2) presentation, (3) required information, (4) general notation, (5) sectioning, (6) dimensioning, (7) marking, (8) reference, and (9) hints in drawing preparation.

A number of the drawings should have a title block, usually pre-printed, in their lower right hand corner. (See figure 2-3.) The following information is recommended for inclusion in the title block:

a. Descriptive title for drawing
b. Title and location of project
c. Architect's name
d. Name, address and telephone number of Precaster
e. Initials of draftsman
f. Date of issuance
g. Job number
h. Initials of drawing checker
i. Scale
j. Number of each sheet
k. Revision block

At the beginning of the discussion of each type of drawing is a box showing mark, size, and scale. Mark refers to Sheet Prefix Mark: (A-1, A-2, etc. for anchor plans or S-1, S-2 for hardware details.) Size refers to recommended minimum drawing sheet size such as 24 x 36 inch. Scale refers to recommended drawing scale such as $1/8'' = 1'0$ or $1' = 1'0$. The scale used should be indicated on drawings.

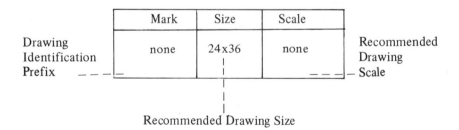

	Mark	Size	Scale	
Drawing Identification Prefix	none	24x36	none	Recommended Drawing Scale

Recommended Drawing Size

Cover Sheet The cover sheet contains all pertinent job information to be used as standard reference for the duration of a project. The material that follows is intended as a guide during preparation. (See figure 2-3.)

Planning. A master cover sheet should be developed containing all of the information in figure 2-3 except the general notes and hand-lettered information. If duplicates are not pre-printed, a sepia should be made from the master sheet and filled out for each job.

JOB DIRECTORY

DRAWING SYMBOLS

GRAPHIC SYMBOLS

ABBREVIATIONS

CONTRACT DOCUMENTS

DOCUMENTS	DATED	LAST REVISED
ARCH DWGS A1 THRU A19	1-1-74	
ARCH DWG A20	1-1-74	2/21/74
STR DWGS S1 THRU S6	1-1-74	
SPECS	1-1-74	
ADD NO 1	2-1-74	

ERECTION DRAWING LIST

NO.	DESCRIPTION
K1	KEY PLAN
E1	SCHEDULE & ELEVATIONS
E2	ELEVATIONS
E3	do
E4	do
E5	do
E6	do
E7	do
E8	ERECTION PLAN
E9	SECTIONS & DETAILS
E10	do
E11	do
E12	do

GENERAL NOTES

1. The Engineer of Record is responsible for the design adequacy, strength and long term behavior of all loose connection hardware shown on approved shop drawings, and location of supporting structure.

2. Precast panels and loose erection hardware only are furnished F.O.B. Jobsite by Architectural Precast Corporation.

3. Special material requirements:
   Concrete f'c = 5000 psi
   MRF f'c = 6000 psi Post insulative paint
   Loose hardware A36

4. In lieu of specific tolerance requirements, tolerances will be in accordance with the "PCI Manual of Quality Control for plants and production of architectural precast concrete products".

5. Responsibility of approvers

6. Architectural Precast Corporation is not responsible for the adequacy for the support structure nor for deviation from support structure conditions as shown by the contract drawings.

7. All inserts on exposed surfaces of precast will be recessed, for patching by others.

8. The erector shall notify Architectural Precast Corporation of shipping damage upon delivery. Lack of notification at unloading shall indicate acceptance of an undamaged precast unit by the erector.

9. Coil bolts used for erection shall have a minimum bolt penetration of 3 in. into the insert.

10. All panels shall be field handled and erected in accordance with the methods and sequences determined by the Erector and approved by Architectural Precast Corporation.

11. All welding shall be done with E60 electrodes. Weld sizes and lengths not noted shall be full length and the size of the fillet shall be 1/16" less than the smallest thickness of material being welded.

12. Finish requirements:

13. Insert Designation:

14. Electrical boxes are to be furnished to Architectural Precast Corporation by the General Contractor prior to casting. These shall arrive complete with all parts integral to the precast such as conduit.

SITE MAP

ARCHITECTURAL PRECAST CORPORATION
20 CONCRETE AVENUE
EVERYTOWN   ILLINOIS
(312)

PROJECT: N.A. MANUFACTURING (S12)
ADDRESS: 1519 OAK ST, HILLVILLE, ILL.
DRAFTSMAN: AL YOUCHIT

Figure 2-3

Presentation. Figure 2-3 illustrates what is generally acknowledged to be the best method of cover sheet presentation.

Required Information:

a. General

The precast company logo should appear along with the name of the chief or project draftsman and the telephone number of the engineering department. The project name, as it appears on the architectural plans, and the Precaster's job number, should also be included on the cover sheet.

b. Telephone Directory

A telephone directory, listing contacts' names, addresses, and telephone numbers should be organized for the following individuals:

(1) Architect (2) Engineer of Record (3) General Contractor's Field and Home Offices (4) Pertinent Sub-contractors and Vendors

c. Site Map

A sketch of the job site indicating major traffic arteries and significant landmarks should be placed on the cover sheet to aid in the location of the site.

d. Notes

(1) Shop—List finish requirements, insert designations, special handling procedures, and stock piling and blocking instructions. Note minimum coil bolt penetrations. (See Appendix L, page 179.) List tolerance requirements, (Appendix E) or make reference to the *PCI Manual of Quality Control for Plants and Production of Architectural Precast Concrete Products.*

(2) Quality Control—List all materials requiring mill tests and certificates of compliance and, in addition, specifications and physical properties of all nonstandard materials (variations from normal production of concrete strength or rebar grade). Specify plant testing by number and frequency of tests.

(3) Erection—Include instructions for return of such items as strong backs and bracing angles. Specify welding requirements and minimum coil bolt penetrations for erection. Request written notification of any damage resulting from shipment immediately upon receipt of panels. Refer the erector to summary sheets for panel handling requirements. Note whether units are furnished erected or F.O.B. Specify contract inclusions.

(4) Miscellaneous—Include notes which define the meaning of approval interpretation by Architect, Engineer of Record, and General Contractor, and list requirements by Precaster of other sub-contractors. Indicate seemingly questionable supports and other unclear conditions. Note what delay in approval will mean in terms of schedules and completion.

e. Contract Drawings

A list of the contract documents used to prepare erection drawings must appear on the cover sheet. Drawing numbers, dates, and latest revisions must be provided. List the architectural and structural drawings, other drawings used, sketches, specifications and addenda. This protects the Precaster from problems arising out of revisions which are never received.

f. Standards

All standard PCI abbreviations, drawing symbols, and graphic symbols provided in this handbook should be pre-printed on the master sheet. And other special symbols and/or abbreviations should also be included.

g. Erection Drawing List

A complete list of all erection drawings should be included on the cover sheet. Sheet number and description should be indicated.

Mark	Size	Scale
K	24x36	1/8 or, proportion

Key Plan

A key plan is recommended for any project requiring more than one sheet of elevations and/or erection plans. This plan gives the reader a common reference, and is especially useful on a project where an extremely complicated building, or number of separate buildings, is involved. A key plan is also useful whenever a general job discussion arises.

The following items are intended as guidelines during preparation of the plan. (See figure 2-4 on following page.)

Presentation. Figure 2-4 provides a good example of key plan presentation. Note that, as the name implies, all buildings are shown in plan view and are drawn as large as the sheet size permits.

Required Information. All major grid lines and dimensions, both intermediate and overall, which relate to precast panels should be indicated. Use an arrow to indicate construction north, and true north if different. (See figure 2-5.)

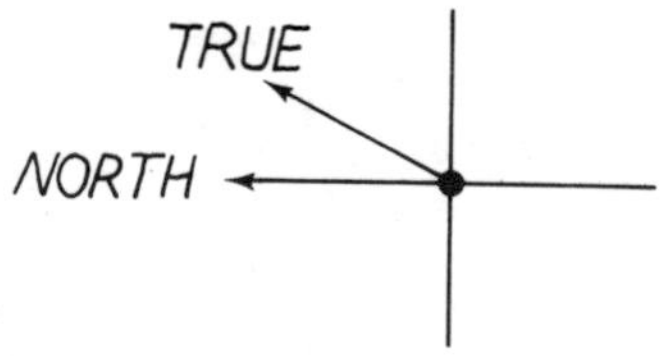

Figure 2-5

Show existing roads, streets, alleys, temporary construction paths, existing buildings, power lines, underground utilities and storage tanks.

Indicate on which sheet each part of the building appears in elevation or plan. Note any variations in vertical height from building to building. Label plan and indicate scale used.

HINTS

a. When space permits, the key plan may be drawn on the cover sheet.
b. If critical, the general erection sequence is shown on the Key Plan.

Mark	Size	Scale
E	24x36	none

Schedules

Schedules are employed for the purpose of grouping together all precast units for ease of reference. Most individuals concerned with a project will have periodic need for this drawing.

Drafting—As they are detailed, panels should be checked off.

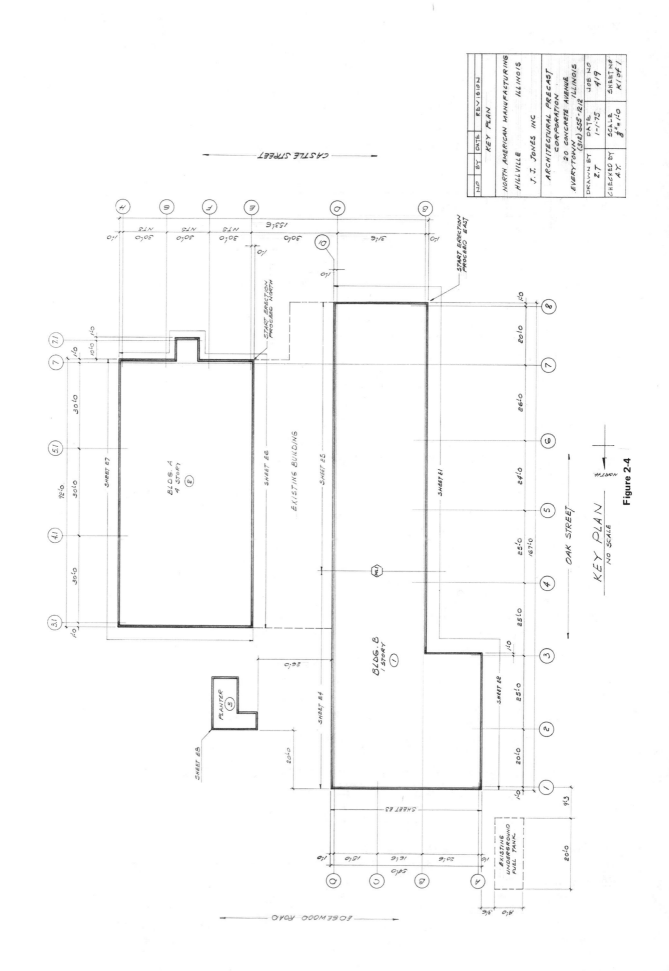

Figure 2-4

Production—Double check quantities, reinforcing cage marks, handling references, and overall size.

Shipping—Schedule loads for maximum size and weight.

Erector—Determine handling sequence for each panel, and check off as they are erected.

A standard pre-printed form (illustrated in figure 2-6) may be used for scheduling information.

PANEL SCHEDULE									
MARK	TICKET	REQD	WT	CAGE	HANDLING	REFERENCE	HEIGHT	WIDTH	DEPTH	REMARKS
F1	T20	20	5250	R20	H1	E1, E2	13'-11½	4'-11½	6"	
F2	T21	1	5250	R20	H1	E2	13'-11½	4'-11½	6"	INSERT VARIATION
F3	T22	1	300	R22	H3	E1	13'-11½	4'-11½	6"	OPENING
F4	T									

Figure 2-6

Mark	Size	Scale
E	24x36	1/8,1/4

Elevations

Elevations are primarily for the use of the erection and shipping departments. Elevation drawings are the Precaster's graphic definition of architectural and structural intent based on all contract documents. They must be clear enough to be checked easily by the Architect, Engineer, and Contractor, and must contain all information vital to the erection process.

When approved, the elevations and sections will replace the contract documents in providing information for production drawing preparation. Along with the other erection drawings, they must contain all the information necessary to produce complete production drawings, and also serve as a reference during shipment and erection. The precast panels will be set from the elevation drawings which must provide the manufacturer with a complete reference in the event questions arise or future building additions or modifications are planned.

The following information is intended as a guideline for use during the preparation of elevation drawings. (See figure 2-7.) Before proceeding it should be determined whether or not panels could be more clearly defined by drawing an erection plan.

Planning. To plan the sheet layout, lightly draw overall outlines of each elevation to eliminate crowding of information. Ample space must be provided for grid lines, dimensions, and notations. Sufficient space should be retained for plot plan, handling schedules, and material lists (if required). In the preparation of drawings it must be remembered that panel identification marks will be added later.

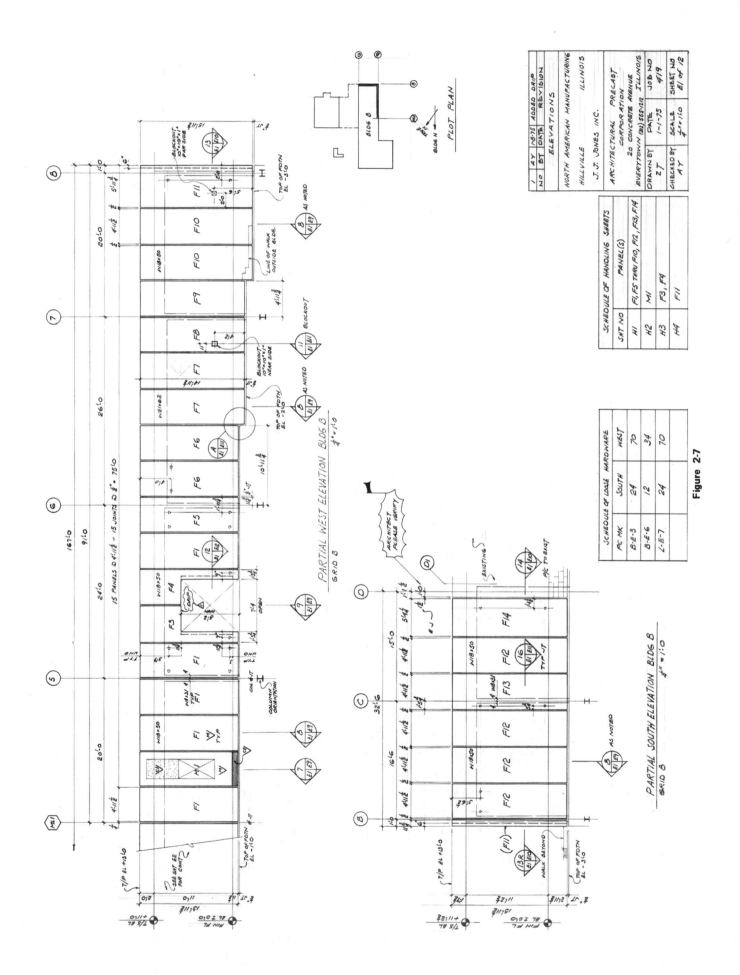

Figure 2-7

Plan the elevations so that final production drawings can be worked out by taking dimensions and details directly from the elevation drawings.

Presentation. Include plot plan with grid lines and true north and/or building north indicators so that the various elevations can be oriented to the entire building. Darken plot plan lines to indicate elevation represented on each sheet. Show all panels, with minimum detail, at least once.

Do not draw the elevation in great detail since sections and shape drawings bear out variations at a larger scale. Label each wall or portion shown with direction and parallel grid line reference as indicated in figure 2-8.

Indicate details in larger scale wherever congested areas prohibit clarity at the normal scale. Do not use opposite hand or multiple reference when labeling elevations. (Refer to figure 2-9 for example.) Elevations should be drawn in one plane of view only. Do not attempt to show staggered elevation views as illustrated in figure 2-10.

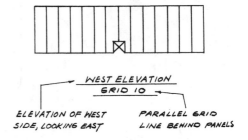

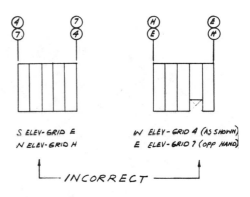

Figure 2-9

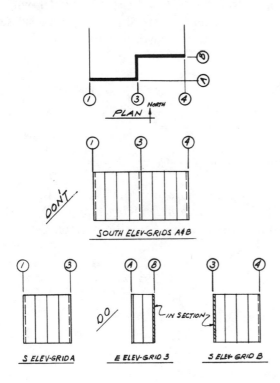

Figure 2-10

Required Information. A well-prepared elevation sheet shows the width of joints, grid lines related to panel joints, and panels related to floor elevations and top of structural support. Incorporate into the drawing all pertinent grid and structure lines. Indicate size and give orientation of any columns or beams located behind or projecting through the precast elevation which may result in the

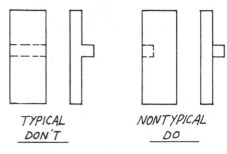

TYPICAL
DON'T

NONTYPICAL
DO

Figure 2-11

need for modification to typical connection insert locations. Identify materials and structures adjacent to precast. Note scale that is used.

All nontypical returns and blockouts on the panel backs should be drawn as dotted lines. (See figure 2-11.) Schematically locate connection inserts and other hardware for each panel.

Include a bill of materials for loose or field cast items supplied by the Precaster, plus a handling schedule.

General Notations. To ensure the insertion of proper notation, refer to the following pages:

Drawing Symbols . . . . Appendix F, page 163

Graphic Symbols . . . . Appendix F, page 164

Abbreviations . . . . Appendix D, page 150

Insert Designations . . . . Appendix F, page 169

Sectioning. Using the 3-key systems, (see Drawing Symbols, page 164), cut sufficient horizontal and vertical sections to completely define the following items for each precast concrete panel:

Shape   Connection

Location in Structure   Blockouts and/or Openings

Finish   Erection Sequence

Joint Arrangement and Sealant

It is often helpful to note any rotated sections as illustrated in figure 2-12. Refer to *Section Drawings* (page 37) in this chapter for additional information on sectioning practices.

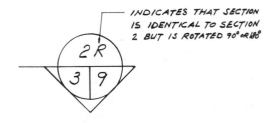

INDICATES THAT SECTION
IS IDENTICAL TO SECTION
2 BUT IS ROTATED 90° OR 180°

2 R

3 | 9

Figure 2-12

Dimensioning. Overall dimensioning should be performed so that each panel height and width is given in elevation. Do not dimension from center line of joint to center line of panel joint. Dimension horizontal and vertical joint widths typically, at expansion joints and corners. Dimensionally tie foundations, floor, and roof levels to top and bottom of panels. List grid to grid and overall dimensions for each elevation and dimension relationships between grid lines and panel edges. Size all blockouts and openings and locate them to a panel edge and horizontal reference line. Indicate width x height x thickness for each opening or blockout. Refer to Appendix K, page 178 for dimensioning of blockouts and openings.

Marking. The mark number for each individual panel must be shown in large print. If the panel appears more than once, the second time its mark should be placed in parenthesis to avoid counting the panel twice. Marks should be kept as short as possible. A letter/number combination for the mark number should be used, as either one alone often becomes lost.

When marking panels make the most common unit the number one unit. Do not mark panels which are mirror images of one another—lefts and rights—as such. (See figure 2-13.) *If computerized accounting is used letter designations may pose a problem.*

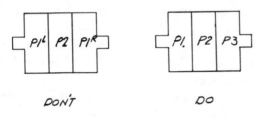

DON'T   DO

Figure 2-13

Reference. To define the panel shape of a complex unit make reference to shape drawings. This will eliminate the repetition of lines already shown on shape drawings. Refer to handling details and anchor plans by means of schedules.

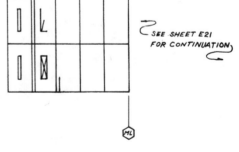

Figure 2-14

HINTS

a. Clearly indicate any dimension or detail that is critical or questionable, and request verification. Indicate who should verify. Call attention to areas which were purposely changed.

b. Section views and details may be included on the elevation drawings for small jobs.

c. Use match lines to break long elevations. Refer to sheet where continuation is shown. (See figure 2-14.)

d. Although it is not recommended, some architectural treatment may be incorporated on the elevations. (See figure 2-15.)

e. Cut erection plans on elevations and make reference to them.

f. When marking panels or assigning grid lines, do not use the letters *I*, *O*, *U*, *V*, and *Z* as they are easily confused with other letters and numbers.

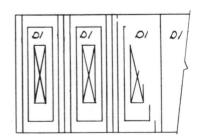

Figure 2-15

Mark	Size	Scale
E	24x36	1/8, 1/4

Erection Plan   Erection plan drawings are normally only used when elevations do not clearly define a specific area. Plan drawings are especially useful for indicating beams, columns, column covers, handrails, planters, and coping.

Erection plan drawings—as elevation drawings—are a means by which the Precaster interprets the architectural and structural plans based on all contract documents. *Upon approval these drawings, together with other erection drawings, replace the contract documents in providing information for production drawing preparation.*

They must contain all of the information necessary to produce complete production drawings, and serve as a reference during erection of the panels. These plans must provide the manufacturer with a suitable future reference should questions arise, or building additions or modifications be required.

The information that follows is intended as a guideline for use during the preparation of erection plans (see figure 2-16).

Planning. Plan sheet layout by lightly drawing the overall grid system. This will ensure that ample space is retained around the plan for grid lines, dimensions, and notations. Space must also be provided for material lists, plot plan, handling schedule, and later, addition of panel marks.

Presentation. Show all panels, in minimum detail, at least once. Indicate a small scale plot plan with exterior grid lines and an arrow indicating building north. Plot plan lines should be darkened to indicate areas represented on each sheet. In the case of an interior building plan, show all precast units in the space at or between the floors represented and the floor above as illustrated in figure 2-17. Do not use opposite hand or multiple reference when labeling plans.

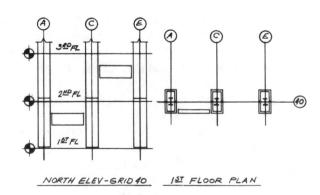

NORTH ELEV-GRID 40   1ST FLOOR PLAN

Figure 2-17

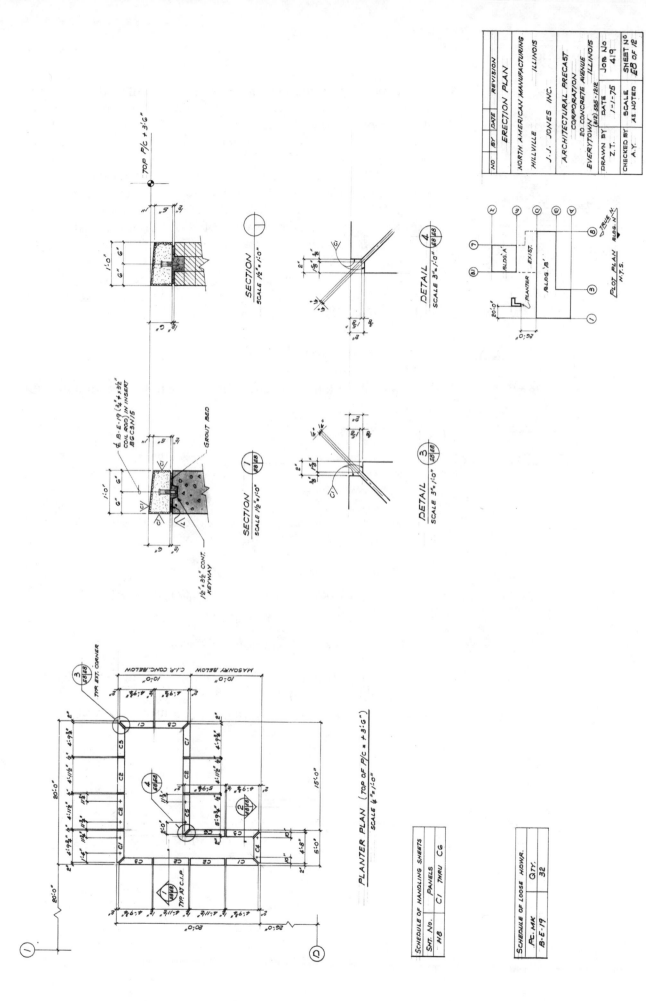

Figure 2-16

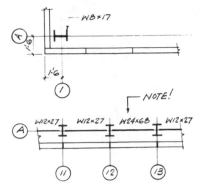

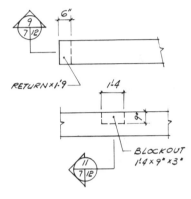

Figure 2-18

Figure 2-19

Required Information. Indicate all pertinent grid lines in the vicinity of the precast panels. Show building north and true north (if different) on the plan. Indicate size and give orientation of any columns and beams which influence connection insert locations. (See figure 2-18.)

Indicate precast supports and note any variations. Returns and blockouts should be shown as dotted lines. (See figure 2-19.)

Materials and structures adjacent to precast should be identified. The scale must also be noted. A bill of materials should be included for all loose or field cast items supplied by the Precaster.

Schematically show anchor locations once for each panel mark. Indicate dimensions from center line of anchor to sides or end of panel. Show vertical location of plan by giving the floor line, roof level, or top of precast elevation. A handling summary table should be included.

General Notation. Use symbols shown in the sections and pages listed:

Drawing Symbols .... Appendix F, page 163
Graphic Symbols .... Appendix F, page 164
Abbreviations .... Appendix D, page 150
Insert Designations .... Appendix F, page 169

Sectioning. Cut adequate vertical and horizontal sections to completely define the following items for each panel:

Shape	Connection
Location	Blockouts and/or Openings
Finish	Erection Sequence

Joint Arrangement and Sealant

It may be helpful to note rotated sections or details. Refer to figure 2-12 (page 32) for example. Details of congested areas should be drawn to a larger scale. Ensure uniform section orientation as shown in figure 2-20.

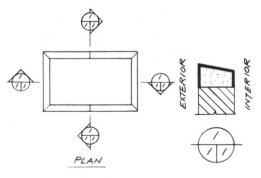

Figure 2-20

Dimensioning. Length and width dimensions must be shown for each type of panel, and overall dimensions of building given for each side of the plan. Dimension joint widths, typically, at expansion joints and corners, as well as distance between grid lines and relationship between grid lines and panel edges. Size all blockouts and openings and relate dimensionally to panel edges. List length × height × width. Refer to Appendix K for method of dimensioning blockouts and openings.

Marking. In large print, place mark numbers on each panel. Make the most common unit the numbers one unit. Do not mark lefts and rights—panels which are mirror images—as such. (See figure 2-21.)

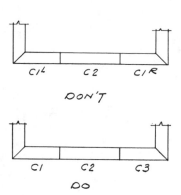

Figure 2-21

Reference. To eliminate repetition of dimensions which are already shown on the shape drawing, and to define the configuration of a complex unit, make references to shape drawings. This eliminates the possibility of an error in copying. Make reference to handling details and to anchor plans for anchor locations.

HINTS

a.   Clearly indicate any detail or dimension that is questionable and/or critical, and request verification. Indicate who is to verify.

b.   When working on small jobs, section views and details may be included on erection drawings.

c.   If a panel is nearly symmetrical to the eye, but can not be turned end for end, place mark number at one end to indicate position at job site. This will correspond to the mark placement on the panel ticket. A second method to ensure proper job site positioning, is to mark one end of panel to correspond with the direction it faces. (See figure 2-22.)

d.   Use match line with number or letter and refer to plot plan if large scale is required for clarity. Refer to sheet where continuation is shown.

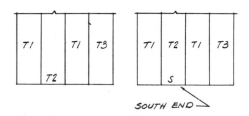

Figure 2-22

Mark	Size	Scale
E	24x36	½,¾,1,1½,3

Section drawings are the final interpretation of the Precaster's elevations or erection plans. The sections are submitted for approval together with elevations or erection plans.

When approved, they replace the contract documents in providing information for production drawing preparation. In conjunction with other erection drawings, the section drawings must contain all the necessary information to produce complete production drawings without reference to contract documents and, in addition, serve as a reference for shipping and erection.

These drawings provide the manufacturer with future reference in the event questions arise or additions or modifications are made to the structure.

The following is intended as a guide to be used during the preparation of section drawings. (See figure 2-23.)

Section Drawings

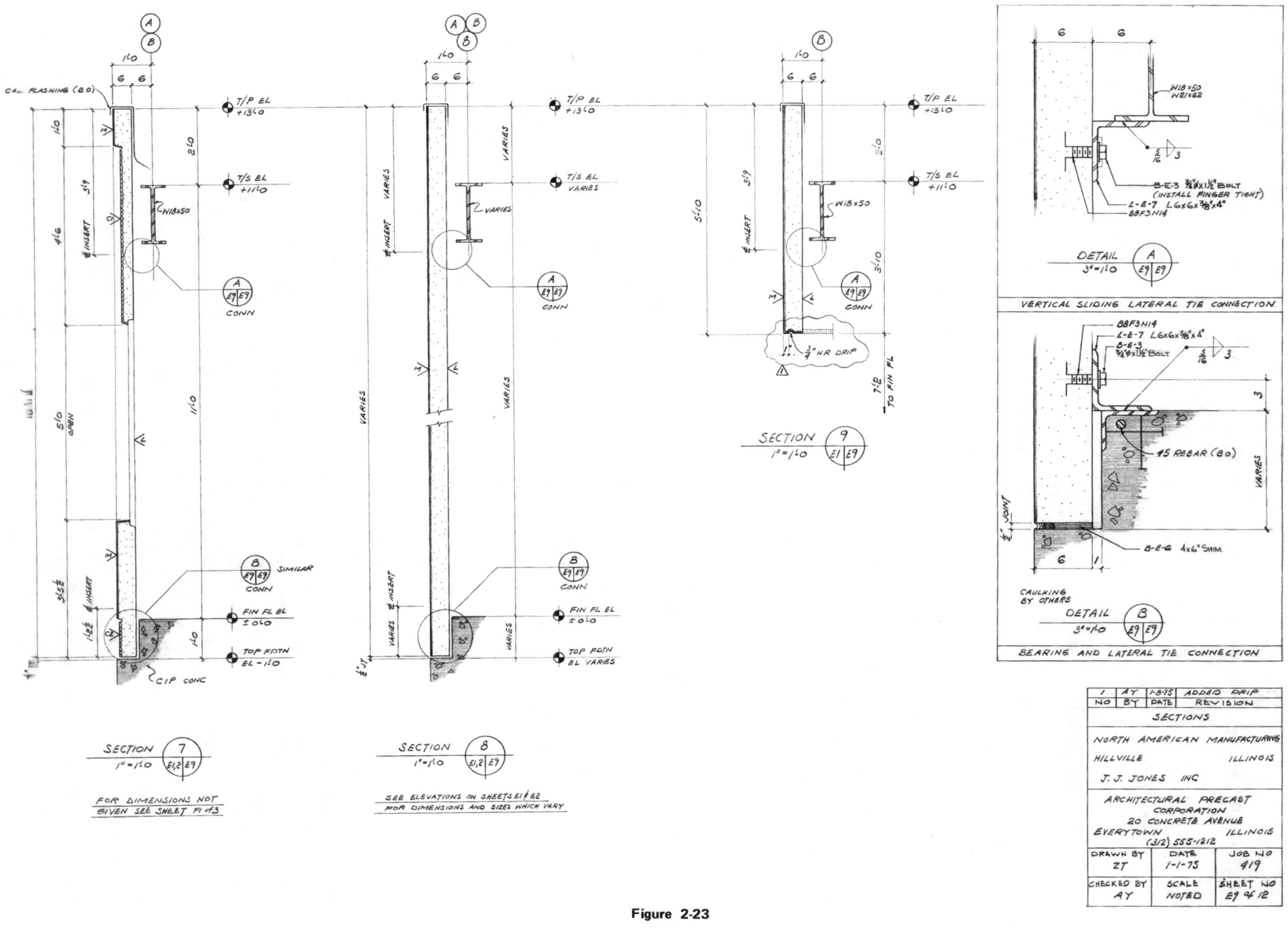
SECTION 7
1" = 1'-0
FOR DIMENSIONS NOT GIVEN SEE SHEET F1 of 3
SECTION 8
1" = 1'-0
SEE ELEVATIONS ON SHEETS E1 & E2 FOR DIMENSIONS AND SIZES WHICH VARY
SECTION 9
1" = 1'-0
T/P EL +13'-0
T/S EL +11'-0
T/S EL VARIES
FIN FL EL ±0'-0
TOP FDTN EL -1'-0
TOP FDTN EL VARIES
C.I.P. CONC
3/4" HR DRIP
TO FIN FL
DETAIL A
3" = 1'-0
VERTICAL SLIDING LATERAL TIE CONNECTION
B-E-3 3/4"x1 1/2" BOLT (INSTALL FINGER TIGHT)
L-E-7 L6x6x3/8"x4"
B8F3N14
DETAIL B
3" = 1'-0
BEARING AND LATERAL TIE CONNECTION
#5 REBAR (BO)
B-E-6 4x6" SHIM
CAULKING BY OTHERS
1 | AY | 1-8-75 | ADDED DRIP
NO | BY | DATE | REVISION
SECTIONS
NORTH AMERICAN MANUFACTURING
HILLVILLE ILLINOIS
J. J. JONES INC
ARCHITECTURAL PRECAST CORPORATION
20 CONCRETE AVENUE
EVERYTOWN ILLINOIS
(312) 555-1212
DRAWN BY ZT | DATE 1-1-75 | JOB NO 419
CHECKED BY AY | SCALE NOTED | SHEET NO E9 of 12

Figure 2-23

Planning. A check print should be made when elevation drawings and/or erection plans have been completed. By use of this check print, the draftsman determines where each section must be cut.

Sections must be cut at the following points:

a. At each different precast cross section.
b. At each significant change in the supporting structure.
c. At openings to show edge requirements.
d. At joints to indicate joint sealants or grouting.
e. At expansion joints.
f. At exterior corners.
g. At interior corners to show return panels.
h. At hardware to show position, size, and spacing.

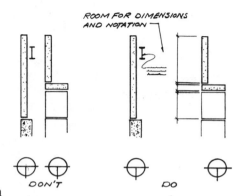

Figure 2-24

The draftsman should provide ample space around each section for all notations which will be required. (See figure 2-24.) Notes should not extend appreciably beyond their respective views or they may distract from it or confuse adjacent views.

Sections are cut in the following manner:

a. Section indicators must be placed on the elevation or erection plans to reflect exactly what is to be seen in each section.
b. Cut lines may be staggered or broken; they need not be continuous. Drawn in this manner, they may include as much information as possible and at the largest scale permissible. (See figure 2-25.)
c. Horizontal sections should be cut looking down.
d. Vertical sections should be cut looking left.

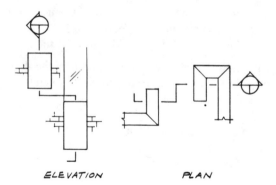

ELEVATION   PLAN

Figure 2-25

Presentation. Horizontal and vertical sections should be grouped on separate sheets. Larger scales must be used if the recommended scale does not adequately show all of the pertinent details and dimensions. The sections must be all-inclusive and should be numerous enough to cover all conditions. It is better to show several sections rather than one confusing section.

Graphics should be used to differentiate between precast concrete and other materials. Areas cut are shown graphically as being in section.

General Notations. To ensure use of the proper symbols, refer to the sections and pages that follow:

Required Information. The required items, or information, may be considered a "skeleton" form of the section. Whenever sections are drawn, the following items should be considered:

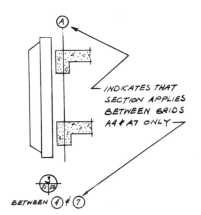

a. Show precast panel.

b. Note finishes on all surfaces.

c. Dimension the following: (1) panel size, (2) panel relationship to structure, including all clearances. (Refer to page 161.)

d. Show and give size of structural supports if they affect the location or function of a connection. (See page 198 for proper nomenclature.)

e. Show architectural surroundings (without going into elaborate detail) if pertinent to attachment or finish so that the relationship of the precast element to the balance of the structure can be seen clearly.

f. Note elevations:

(1) Top of structural support

(2) Top and bottom of precast panel

(3) Grade

g. Identify grid lines running parallel to precast panel. Note perpendicular grid or refer to elevation plan. (See figure 2-26.)

Figure 2-26

h. Indicate mark, (see page 50) type and location of the following hardware:

(1) *S* type hardware (hardware cast into panels in the shop)

(2) *C* type hardware (hardware in cast-in-place concrete)

(3) *E* type hardware (erection hardware)

(a) Angles

(b) Bolts (give diameter and length)

(c) Washers

(d) Shims (give size and type of material)

(e) Dowels

(f) Nuts

(g) Loose plates

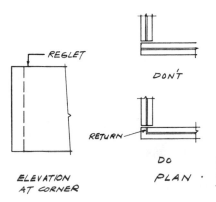

ELEVATION
AT CORNER

i. Dimension joints and specify material used to seal them.

j. List size, type, and location for all flashing reglets and nailers. Watch for returns. (See figure 2-27.)

Figure 2-27

k. Locate all required welds and give size and length.

l. To avoid duplicating information already presented, make reference to other appropriate precast drawings such as:

(1) Anchor plans for lateral location of cast-in anchors.

(2) Shape drawings for intricate detail and panel dimensions.

(3) Elevations for exact locations where section applies.

(4) Similar details in other section drawings.

m. Indicate scale used.

n. Coordinate section using the 3-key system. (See page 164.)

o. Give size, location (see Appendix K, page178), and function of all openings.

p. Size and locate reinforcement, if required (this is optional and done only when specified by the Architect or Engineer of Record). If reinforcement is elaborate a separate sheet may be used.

q. Give functions of all connections, i.e. bearing, lateral, etc.

If a shortened view is necessary use breaklines and note all major dimensions not to scale. (See figure 2-28.)

When cutting a section for a multi-story building, use break lines to show typical sections from floor line to floor line, and note which floors the "typical" section represents. (See figure 2-29.)

Where necessary, show enlarged details to clarify congested areas too small to be shown at normal scale. Details that normally require enlargement are drips, reveals, complicated connections, window jambs and sills, and small radii. (See figure 2-30.) Enlarged details can also be shown in isometric or oblique views to add clarity to a drawing.

Dimensioning. The following list of dimensions must be clearly shown for each type of section. The draftsman must ensure that all secondary strings of dimensions total the overall string.

a. Overall panel size.

b. Relationship between panel and structure. All sections must be coordinated with the elevations by dimensions to column and floor lines.

c. Location of any cast-in hardware other than that required for production and erection handling.

d. Joint width.

e. Finish limitations.

f. Distances between precast and any obstacle which may present erection problems. This can be an existing building or structural member that will interfere with crane lines.

Dimensions which must be held to a tight tolerance should be noted as critical. If the tolerance is known, it should be listed. (See figure 2-31.)

It should be re-emphasized that all intermediate dimensions must add up to the overall dimension. A section is not complete until this has been accomplished. When dimensions are applied to a drawing, the draftsman should determine their usefulness by asking himself if he could execute all details with the dimensions provided.

Reference Drawings. Drawings used by the draftsman during the preparation of sections frequently consist of more than the standard architectural and structural contract drawings. He also should refer to the following drawings in the preparation of sections.

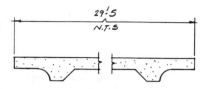

Figure 2-28

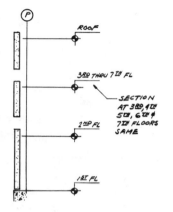

Figure 2-29

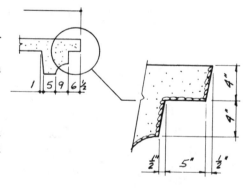

Figure 2-30

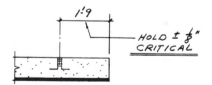

Figure 2-31

(1) Mechanical and electrical drawings to determine hole and fixture requirements. These requirements are normally incorporated by the General Contractor when checking Erection Drawings for approval.

(2) Structural steel shop drawings for changes from structural contract drawings.

(3) Approved anchor plans for location of anchors cast-in or pre-welded to the structure.

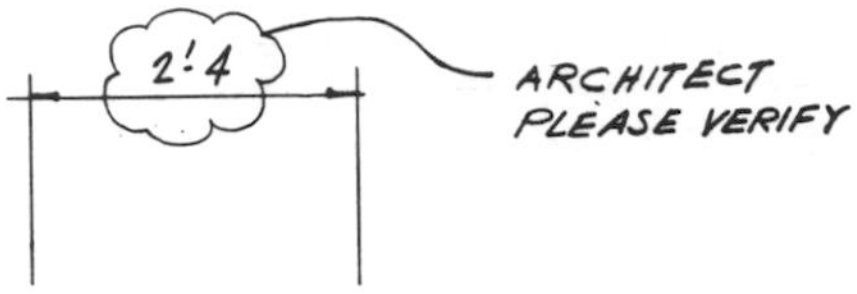

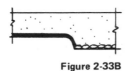

Figure 2-32

HINTS

a. Place a cloud on the back of sheet around any dimension that requires verification. Indicate who is to verify. (See figure 2-32.) Placing it on the back allows future removal without reworking adjacent lines or lettering.

b. When determining finishes, always specify a smooth, flat finish where a loose angle abuts the back of a panel.

Figure 2-33A

c. When two different types of mixes or finishes meet on an exposed panel or surface, indicate a demarcation feature as shown in figure 2-33(a). If a demarcation groove occurs near a change of section, it may create a weakness and counter any attempt to provide a gradual transition from one mass to another. Consequently, solution B figure 2-33(b), is normally preferable, or the thickness of the unit must be changed to compensate for the groove.

Figure 2-33B

d. When several panels are shown in one section, provide an erection sequence if needed. (See figure 2-34.) If more than one set of sequences occurs on the same sheet, group them into sets. (See figure 2-35.)

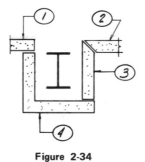

Figure 2-34

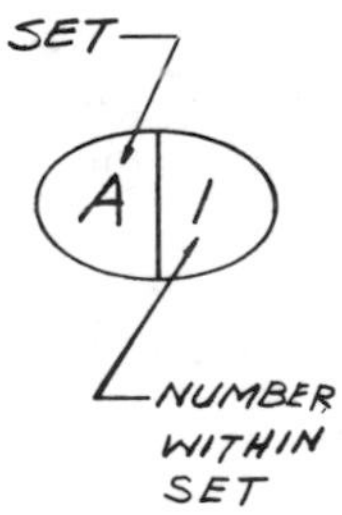

Figure 2-35

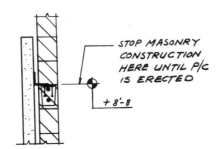

Figure 2-36

e. Indicate erection restrictions posed by the structural materials if required. (See figure 2-36.)

f. Indicate any anchors or inserts which must be recessed for patching. Indicate when, and by whom, patching is to be completed.

g. Avoid reduction in panel sections. (See figure 2-37.)

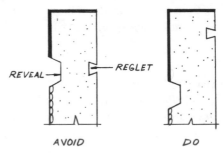

Figure 2-37

h. Use of gray cast iron inserts in precast for structural connections are not recommended due to their brittle nature. Malleable cast iron may be satisfactory.

i. Do not use channel type (continuously slotted) inserts in precast connections unless carefully studied and evaluated by the Precast Engineer.

j. Include notation of possible corrosion protection for details of lifting devices where such hardware is left in the units. If the handling devices interfere with any other function, drawings should include removal instructions as well as possible field repairs.

k. Do not specify field welding of galvanized material. Even if painted afterwards, the galvanized effect is lost in the back of the weld where it is difficult for the paint to be applied.

l. Do not place dissimilar metals in contact with each other unless experience has shown that no detrimental galvanic action will occur.

m. Avoid casting aluminum into precast. If you must, be sure that it is electrically insulated by a permanent coating such as bituminous paint, alkali-resistant lacquer, or zinc chromate paint.

n. Avoid casting wood (nailers, lath) in precast. If unavoidable use short pieces evenly spaced, rather than one continous piece, and bevel sides (reverse draft). (See figure 2-38.) The most suitable wood is pine or fir with a high resinous content.

o. Call out number of connections per panel if required for clarity, or refer to anchor plan. (See figure 2-39.)

p. Standardize on a minimum number of hardware types. Use heavier hardware at lightly loaded conditions if it will eliminate a special piece mark and result in only slightly more material. This practice prevents the possibility of the smaller piece being used where the heavy one is required. (See figure 2-40.)

q. When detailing erection angles for bolted connections, always consider washer size and "k" distance (see *AISC Manual*-Appendix A). Check for interference.

r. Minimum size for steel stock used for connections should be ⅜ inch.

s. Threaded rod material properties vary. Check with the Precast Engineer.

t. If connection hardware must be galvanized, do so following fabrication.

u. Check all connections for proper initial and long term tolerances. (See page 161.)

v. When approval is required for reinforcement, it should be indicated on the section drawings. This practice, however, is not recommended.

w. Note any material shown but not furnished by the precast manufacturer as being "by others".

x. Double bill sections only when variations are minor. (See figure 2-41.)

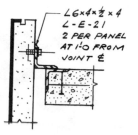

Figure 2-38

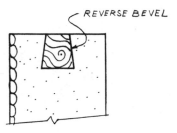

Figure 2-39

If design indicates these sizes for a particular project   ...standardize on these sizes

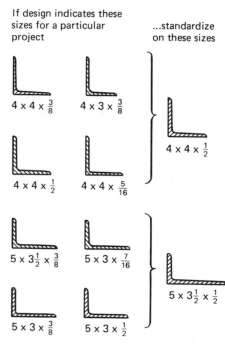

Figure 2-40

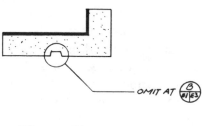

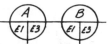

Figure 2-41

y. If welding is critical, specify certified welder only.

z. With bolted connections, ¾ inch or 1 inch diameter bolts are considered standard in the precast industry and should be used. Occasionally 1¼ inch diameter bolts are needed. Regardless of the load requirements, a ¾ inch diameter bolt should be the minimum size used for any precast concrete connection. It is equally important to consider the types of threads being used in bolted connections and to select those that are considered to be standard.

Mark	Size	Scale
H	8½ x 11†	proportion

†24 x 36 quadranted

Handling Detail   A handling detail conveys to shop personnel and to the erector the handling procedures that have been selected by the Precast Engineer. It spells out, in sequence, all procedures for a given panel or group of panels.

The section that follows is intended for use as a guideline during the draftsman's preparation of the handling details. (See figure 2-42.)

Planning. Determine the number of handling procedures for each panel in question (see Appendix P, page 183.) Divide the detail sheet into an equal number of parts. Provide space for a supplementary view of the erection procedure, if needed.

Presentation. A good example of handling detail presentation may be found in figure 2-42. Since it is recommended that panel thickness be exaggerated for purposes of clarity, it is not necessary that handling details be drawn to scale. Note the horizontal lines which separate each step.

Required Information. The following procedures must be illustrated in the drawing:

a.   Stripping

b.   Turning (if required)

c.   Temporary blocking (if required)

d.   Storage

e.   Shipping

f.   Erection

g.   Shoring (if required)

Provide panel marks applicable to the summary. Show and locate all blocking points. Show, dimensionally, which inserts are used for each operation. Give location and type of all dunnage and/or padding required. Indicate stacking orientation for storage and erection, and show all crane lines. Give angles of crane lines if critical to handling.

To avoid preparation of handling details it is sometimes practical to use the code shown on pages 183-86 for schedule drawing and panel tickets, if procedures are standard. (See figure 2-43.)

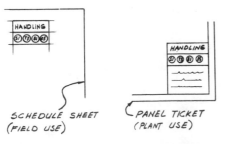

SCHEDULE SHEET (FIELD USE)   PANEL TICKET (PLANT USE)

Figure 2-43

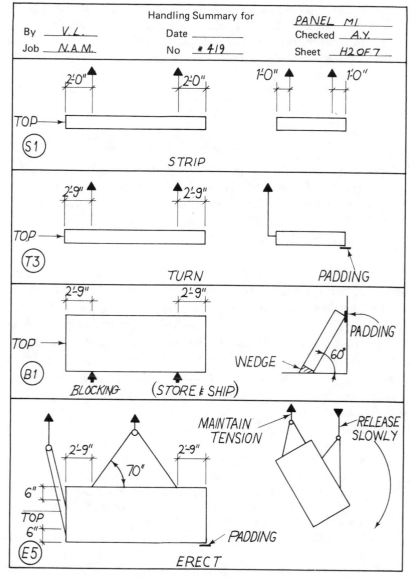

Figure 2-42

Note variations in blocking symbol depending on procedures involved (see page 163.)

HINTS

a. A basic axiom to storage and shipping is to support the precast unit at two points* only. If support is continuous across three or more points, precautions must be taken so that the panel will not bridge over one of the supports (because of differential support movement) and result in bowing and cracking.

b. Often the manner of storing depends on how the panel is to be shipped and what limitations the panel's cross sections impose on handling. For all practical purposes, the panel should be stored in the same manner in which it will be shipped.

*Relative to blocking, this is an industry term denoting a line of support.

c. Most precast concrete firms use either flatbed or low-boy trailers to ship panels, and these units suffer excessive distortions during hauling. Thus, support at more than two points on a trailer unit can be achieved only after considerable modification of the trailer.

Mark	Size	Scale
A	24x36	1/8,1/4

Anchor Plan

Anchor plans generally constitute the first direct contact between the Precaster and job site personnel. (See page 85.) These are the plans used to indicate the location of hardware—either cast-in or pre-welded—to the structure for the purpose of fastening the precast concrete. (See figure 2-44.)

Planning. While anchor plans are relatively simple to prepare, they should not be taken for granted. The best method of showing anchor locations is to draw one anchor plan for each structural elevation or floor line (unless plans for more than one floor are identical) and draw sections for the different conditions. Sections are necessary to adequately define and show the position of the anchors. If anchor locations vary only slightly from one level to another, a master plan can be drawn, and the other floors completed from sepias. (*Note*: The use of sepias is covered in detail beginning on page 9.) Allow space for a material list.

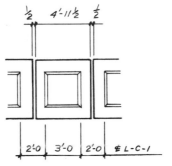

Figure 2-45

> It is often possible to provide anchor locations on the elevation sheets and refer to the sections for the location of bolts or plates in the structure, thus eliminating the need for an anchor plan. (See figure 2-45.)

Basic Presentation. Anchor locations are indicated in plan view. When plan is cut between floor levels, all of the structure appearing in section is shown by means of standard graphic notations. (See figure 2-46.)

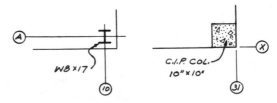

Figure 2-46

Cast-in or pre-welded hardware appearing at the top of the level shown should be indicated with solid lines. All hardware at the underside of the structural level should be shown as dotted lines. Avoid drawing reflected plans.

Any hardware in a column or wall above the floor plan drawn (up to the underside of the next level) should be covered on this

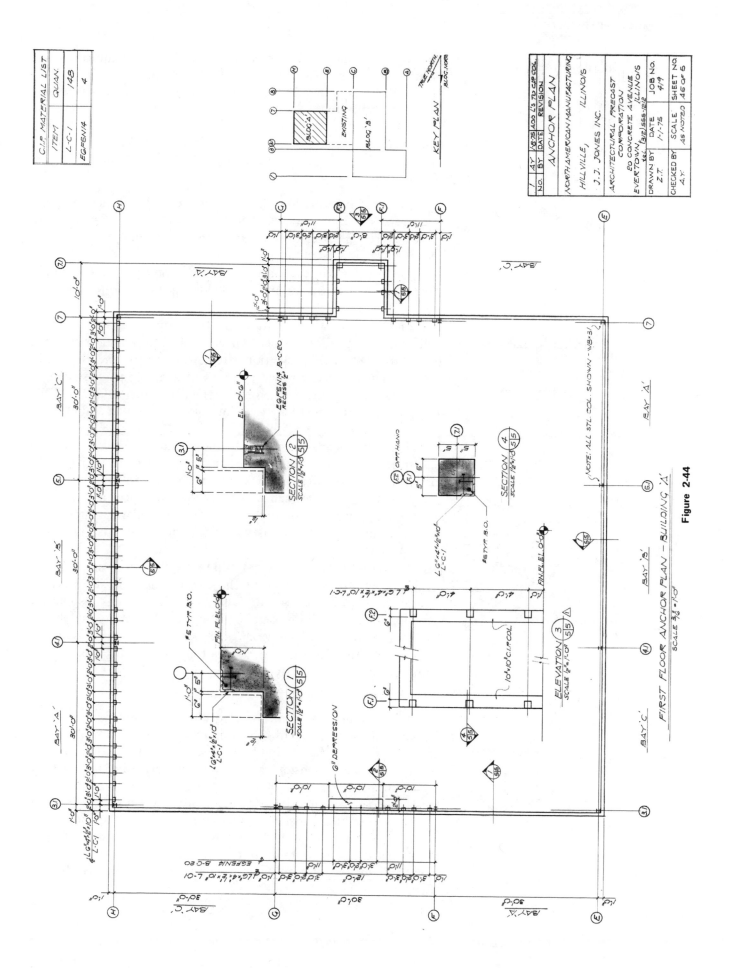

Figure 2-44

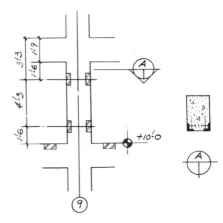

Figure 2-47

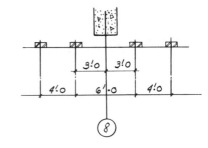

Figure 2-48

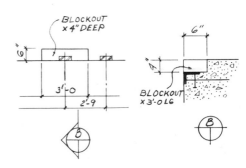

Figure 2-49

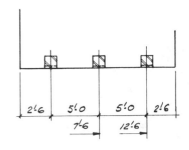

Figure 2-50

plan, and a supplementary elevation and/or section indicating the location should be used for clarification. (See figure 2-47.)

List the size of all structural components to which precast will be connected. Structure in immediate area should be adequately defined and dimensioned. All pertinent grid and column lines must be shown. The sections can be drawn in vacant areas on the sheet or on a separate sheet.

Required Information. List job name and number and the scale used on each drawing. Label the level or floor shown and indicate true as well as building north. Include plot plan and indicate area represented.

Tie anchors into adjacent column and grid lines with a dimension. (See figure 2-48.) As always all dimensions should be given in feet and inches.

Indicate size and elevation of structural components which will receive the precast panels. Give size, location, and orientation of columns when their location affects the precast anchor layout.

All depressions or raised portions of a cast-in-place structural frame should be shown when they affect the anchor layout. Their location and size must be given dimensionally in plan and clarified in a section. (See figure 2-49.)

The type of anchor should be indicated in plan. Use standard hardware nomenclature for plates and angles. (Refer to page 50.) Use standard insert designations. (Refer to page 169.)

Cut a section through each type of hardware, and through different applications of each type. This will clarify dimensions and structural configurations which cannot be adequately shown in plan view.

In cast-in-place concrete framing applications, show rebars which occur in the vicinity of the cast-in anchors. Make certain that anchors do not cause interference. If Precaster is responsible for supplying the cast-in-place hardware, a bill of materials should list the quantities of material. Do not specify other sub-contractors if the information is not included in the precast contract—simply note "By Others".

Dimensioning. The anchor plan must be dimensioned so that workmen are not required to add or subtract dimensions to locate anchors. It may be desirable to use continuous dimensioning (supplementary dimension) as hardware is normally laid out from a grid line with a 50 to 100 ft. tape (figure 2-50). Provide overall and grid-to-grid dimensions for each plan. Always dimension to the centerline of an anchor. Dimensionally tie each anchor to the adjacent grid line or accessible reference point, such as the face of cast-in-place concrete. Indicate dimensions on one or more typical bays, and make notation that others are the same, if applicable. (See figure 2-44.)

Dimensions which must be held to close tolerances should be noted as critical and the tolerance should be listed. Do not dimension different types of anchors in the same string of dimensions. (See figure 2-51.)

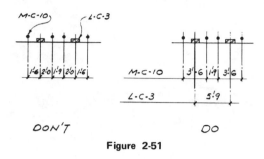

Figure 2-51

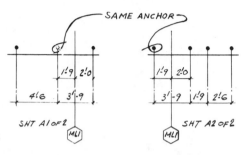

Figure 2-52

HINTS

a. Overlap dimensions and match lines on all sheets where match lines appear. Do not cut dimension lines with match lines. (See figure 2-52.)

b. Ensure that anchors provide adequate initial and long term tolerances. (Refer to page 161 for additional information.)

c. Although not recommended, if anchors must be pre-welded to structural steel, they should be set back from the planned location of the precast panel to accommodate allowable variations in panel thickness and placement. (See figure 2-53.) *Consult Precast Engineer if this condition occurs.* Check to verify that realistic tolerances can be achieved. (See Appendix E, page 155.)

d. When drawing a partial anchor plan, indicate on plot plan which portion of the level is shown.

e. Anchors in cast-in-place concrete columns should be shown by supplementary elevation. (See figure 2-47.) Dimension from floor line to floor line.

f. When a beam steps within a bay, or contains a depression, an elevation of that bay serves to clarify a number of conditions. Sections also should be cut through such areas. (See figure 2-54.)

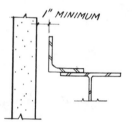

Figure 2-53

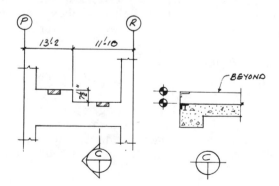

Figure 2-54

g. Locate anchors to allow the greatest amount of repetition in precast panels.

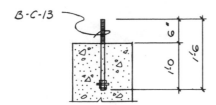

Figure 2-55

h. Dimension the embedded length of anchor bolt when used. Also give length of bolt projecting from concrete. (See figure 2-55.) Specify the use of double templates if plumbness is critical. (See figure 2-56.)

i. It is sometimes helpful to show dotted panel outline to indicate the relationship between the panel and its anchor.

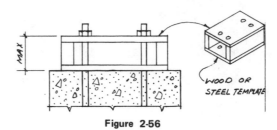

Figure 2-56

Mark	Size	Scale
S	8½x11†	proportion

† 24 x 36 quadranted

Hardware Detail   Hardware details, whether fabricated by a vendor or by the precast manufacturer, must convey all of the information required to manufacture the piece without the need for consulting other drawings. Dimensions, material, method of fabrication, and finish must be shown so that the hardware can be fabricated without encountering questionable details or dimensions. (See figure 2-57.)

Nomenclature. Hardware references should consist of a three-digit sequence. The first digit describes the applicable item. This nomenclature follows the size of the material.

PL—a sheared flat plate.
L—hardware cut from standard rolled angle.
M—hardware fabricated from anything except sheared flat plate or standard rolled angle, or consisting of a combination of shapes. *M* hardware will include bent plates, fabricated wide flanges, channels, and pipe sleeves.
B—bolts, coil rods, washers, threaded rods, shim stacks, nuts, and drilled-in inserts.

The second digit refers to the destination of the hardware.

S—hardware cast into panels, (plant hardware).
E—erection hardware.
C—hardware in cast-in-place concrete or pre-welded to steel in field, (Contractor's hardware).

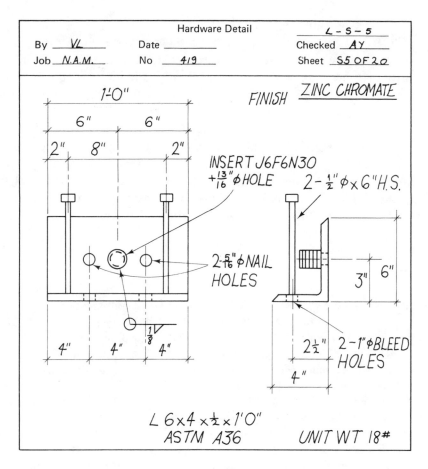

Hardware Detail		$L - S - 5$
By _VL_	Date _____	Checked _AY_
Job _N.A.M._	No _419_	Sheet _S5 OF 20_

FINISH _ZINC CHROMATE_

1'-0"

6" 6"

2" 8" 2"

INSERT J6F6N30
+$\frac{13}{16}$"ϕ HOLE

2-$\frac{1}{2}$"ϕ × 6" H.S.

2-$\frac{5}{16}$"ϕ NAIL
HOLES

3" 6"

4" 4" 4"

$\frac{1}{8}$

2$\frac{1}{2}$" 2-1"ϕ BLEED
HOLES

4"

L 6×4×$\frac{1}{2}$×1'0"
ASTM A36

UNIT WT 18#

Figure 2-57

The third digit is the mark number of the hardware. All mark numbers should start at "1" for plant, erection, and cast-in-place hardware.

Examples

4x4x1/2x6	L-S-3	An angle to be cast into a panel in the shop, and the third shop angle to be detailed.
3x9x3/8	PL-E-5	The fifth erection item detailed—a plate.

General Notations. For the proper notations, refer to the following sections and pages:

Presentation. Hardware details should be drawn on individual 8½- by 11-inch sheets or on large sheets divided into quadrants. Hardware need not be drawn to scale, but should be properly proportioned. Two views are generally adequate to convey the necessary information. Figure 2-57 provides an example of a hardware detail presentation.

Requirements. Include job name and number and individual piece mark. Give overall dimensions of material, size and type of anchors, size of holes and finish. Define the type of materials used, using AISC steel shape designations (Appendix U), and give welding instructions. Indicate the types of inserts used. List unit weights.

HINTS

a. An isometric view will sometimes clarify unusual configurations.

b. Master copies of commonly used hardware items should be kept on file. To save drawing time, make a sepia and fill in the dimensions.

c. The use of fade-out grid paper enables the draftsman to construct a reasonably good freehand drawing using grids as a guide for horizontal and vertical linings. (See page 9.)

d. Size all holes; bolt diameter plus 1/16 inch for bolts to 7/8 inch diameter, and bolt diameter plus 3/16 inch for bolts 1 inch diameter and up.

e. Galvanizing, plating, or painting, when required, should be done following fabrication.

f. Locate holes and anchors on channels, and angles from the heel of the section. If critical, locate holes and anchors in the plates from an established centerline.

g. Slots in erection hardware should be a minimum of 2½ inches long.

h. Anchors must have proper cover, and not interfere with reinforcing steel.

i. Avoid using rebar for anchors. Use headed studs and deformed anchor studs whenever possible. (See figure 2-58.) If rebar anchors are required, do not weld within eight bar diameters of a bend. (See Appendix F for discussion on welding precautions.)

j. Provide holes for attachment to form as well as for release of entrapped air (nail holes, bleed holes).

k. Stagger placement of anchor studs to permit placement of cast-in-place hardware around rebar. Use nut and bolt as an alternate solution. (See figure 2-59.)

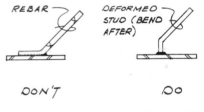

Figure 2-58

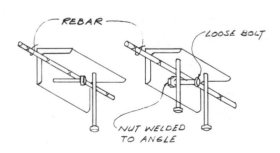

Figure 2-59

Mark	Size	Scale
F	24 x 36	¾, 1

When an architectural precast panel has a complex configuration, it is necesary to submit, for approval, a drawing which thoroughly defines the mold or form shape. If the panel configuration is relatively simple the shape drawing and the panel ticket may be combined. A shape drawing includes only the information needed to build the basic mold, including any major blockouts, such as windows.

The information that follows is intended as a guide during the preparation of shape drawings. (Refer to figure 2-60 on following page.)

Planning. Any panel shape that cannot be achieved by minor modification to another form requires a separate shape drawing. The draftsman can sometimes minimize the number of forms required by means of proper form planning. He should understand the basic mold concepts such as master mold and envelope mold.

The master mold concept is based upon fabricating one master mold (with its appropriate additional tooling) which allows a maximum number of casts per project. Units cast in this mold need not be identical provided the changes in the units can be accomplished as pre-engineered mold modifications.

Modifications should be achieved with a minimum of idle production time and without jeopardizing the usefulness or quality of the original mold. An illustration is provided in figure 2-61.

Shape Drawing

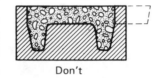
Basic mold to be cut and enlarged for bigger panel

Don't

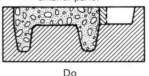
Temporary bulkhead for smaller panel

Do

Figure 2-61

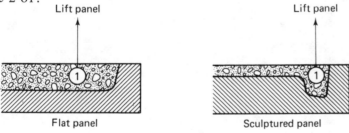

Figure 2-62

A blockout is a type of modification that is relatively easy to accomplish. Bulkheads may be used to modify the precast unit. Such modifications may be used to provide clearance for columns in specific locations, or for precast units which are identical except for window openings. Simple modifications are also accomplished by the use of bulkheads for units which comprise only a part of the master mold.

The complete *envelope mold* is a box mold where all sides remain in place during the entire casting and stripping cycle. Such molds have good economy and quality potential because they are usually simple to build and maintain, with enough strength to withstand concrete pressure during and after consolidation. Figure 2-62 shows a complete envelope mold, while figure 2-63 illustrates

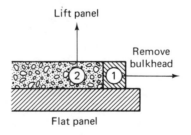
Flat panel

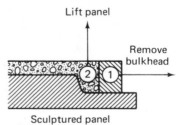
Sculptured panel

Figure 2-63

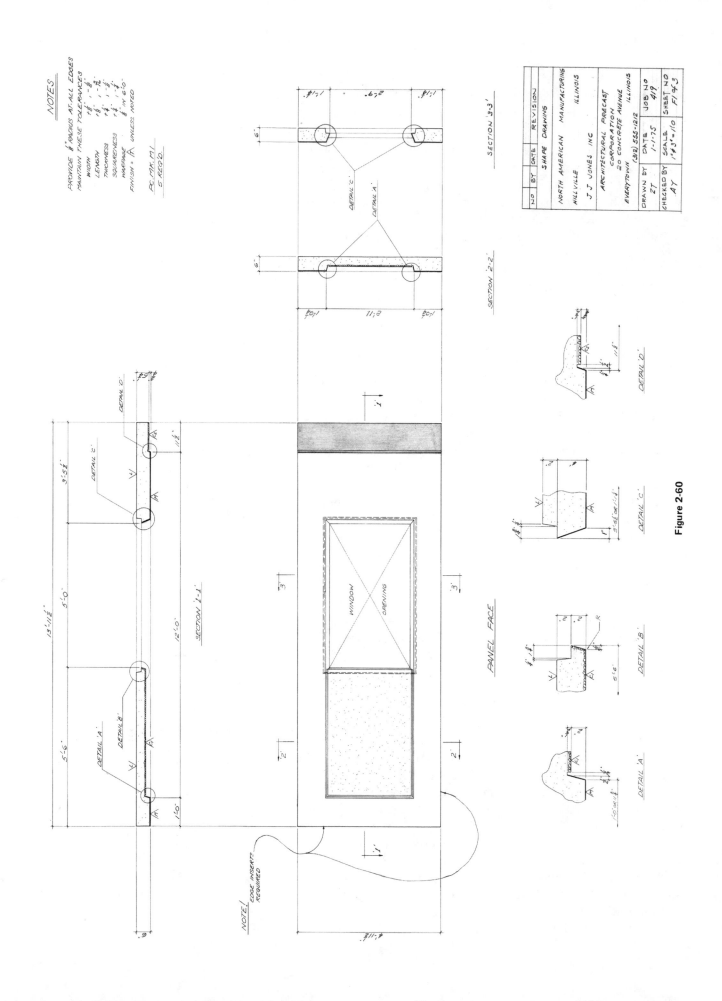

Figure 2-60

the more conventional mold with removable side and end bulkheads. A modification of the complete envelope which will accommodate precast units without drafts along one or more edges is illustrated in figure 2-64.

This mold provides good corner details, but loose side rails are stripped with the unit. The side rails allow 90° returns or returns with negative draft, and are fairly easy to reassemble, eliminating the need for daily measuring and aligning. Such modified envelope molds, however desirable for quality and daily cost savings, cannot justify their initial cost unless reasonable repetition exists. *The draftsman need not concern himself with the number of forms or molds that will be required.* This assessment is made by the production department, however, the draftsman frequently should discuss mold construction with production in order to develop an awareness of cost considerations.

Adequate drafts must be provided on all vertical surfaces to facilitate stripping the precast unit from the mold and to maintain good finishes. Drafts should be shown on the drawing even if it means exaggerating the drawing. This item is often neglected during design, and therefore it may be necessary to call the Architect to resolve a point of aesthetic value. The minimum drafts generally accepted are shown in figure 2-65. Generally, the minimum draft which will enable a unit to be stripped easily from a mold is 1 inch in 1 foot (1:12). This draft should be increased for narrower sections or delicate units as the suction between the unit and the mold then becomes a major factor in both design and

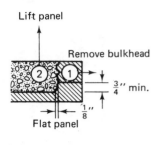

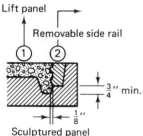

Figure 2-64

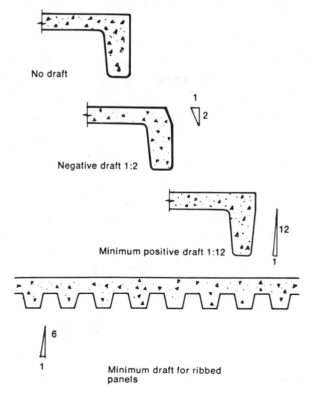

No draft

Negative draft 1:2

Minimum positive draft 1:12

Minimum draft for ribbed panels

Figure 2-65

stripping. The draft should be increased to 1.6 for screen units with many openings, or for ribbed panels.

The drafts selected for finish consideration will vary due to shapes and techniques proposed for production. Vertical sides or reverse (negative) drafts will create entrapped air voids which, if exposed, may be objectionable.

The transition from one mass of concrete to another within a precast unit is a prime consideration. Wherever possible, the transition should be gradual, such as a large radius or fillet in the transition of a mullion to a flat portion in the element. This is necessary to reduce the possibility of cracking and will enhance the structural integrity and the finish of the precast unit. (Refer to the illustration in figure 2-66.)

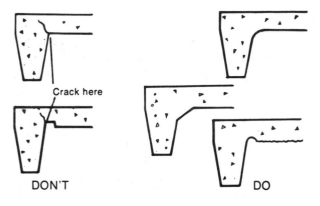

Figure 2-66

Presentation. Figure 2-60 depicts the best arrangement of sections and enlarged details required in the presentation of shape drawings. The primary view is the panel face. The back of the panel is not shown, unless it too is formed.

Required Information. Allowable tolerances for length, thickness, width, and squareness must be provided on the drawing. (See page 158.)

To enable the production department to determine manpower and delivery schedules, the draftsman should indicate the quantity of casts, plus finish requirements, which will help in the selection of the materials to be employed in form construction. Among the materials used are wood, fiberglass, concrete, and steel.

General Notations. Notations used in the preparation of shape drawings may be found on the following pages:

Sections. A separate section is required whenever there is a change in a panel shape. This applies horizontally as well as vertically. As indicated in figure 2-67, each section should be positioned so that it can be read behind the cutting plane.

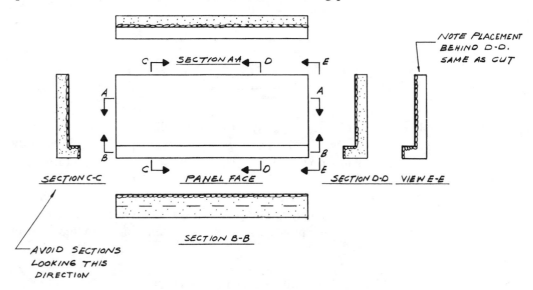

Figure 2-67

Note the type of section indicators used as contrasted to the 3-key system employed elsewhere. Each of the sections must clearly indicate the following:

Dimensions	Portion Beyond (if any)
Finishes	Shape
Portion Cut	

Dimensioning. For ease of reference place overall length, width, and thickness dimensions on the dimension line furthest from the object. Drafts, changes in plane, blockouts, holes, and chamfers must be dimensioned and placed on secondary dimension lines. *These lines are carried full length and tied dimensionally to an edge of concrete easily located from back.*

Figure 2-68

HINTS

a. If a panel is prestressed, show location of strands so that form construction will not interfere. Strand patterns should be standardized.
b. Indicate location of edge inserts or any projecting item which might affect form construction.
c. Show radius or chamfer on any outside corner which will have an exposed aggregate finish. This minimizes form leakage and does not impair finishes. (See figure 2-68.) A minimum radius or chamfer of 1/8 inch will also prevent ragged edges on smooth-finish surfaces.
d. Where possible, and to save time, lay out form drawing so that it may later be converted to a panel ticket. (See figure 2-69.)

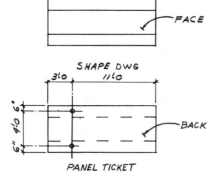

Figure 2-69

e. Note and request verification of any changes in shape from that shown in the contract documents. This includes addition of stripping drafts.

f. Since completion of forms requires the greatest lead time, submit shape drawings for approval at the earliest possible date.

Mark	Size	Scale
T	18 x 24 24 x 36	¾, 1

Panel Ticket

Panel tickets, are the only drawings utilized by the production shop for casting and finishing. Panel tickets are normally only internal plant documents. They must be clear, accurate, and contain all of the information, such as dimensions and details pertaining to the panel, necessary to produce panels which will meet job requirements. The panel ticket serves as a cross reference to shape, reinforcement, hardware, and handling drawings. (Refer to figure 2-70.)

Planning. Each different panel mark should be represented on a separate panel ticket. The panel should be drawn as large as possible, but not necessarily to scale. Sufficient space should be allotted for at least two sections, each describing cross-sectional changes. Space for material lists must be provided. (See Appendix V for required information.)

Presentation. Each panel should be presented in plan view, in the position from which shop personnel will view it. This is illustrated in figure 2-70. Sections and changes in cross section are also shown on this sheet. Sufficient views are necessary to locate all items cast into the unit and to adequately denote the panel shape. The view and sections of the panel are drawn with heavy lines, and clearly labeled. See Hints (page 62) for discussion on the use of schedule (tabular) or schematic methods of presentation.

Required Information. All general panel information—mark number, total number of panels required, panel weight and cubic feet of concrete required—should be in large print so as to be clear and legible to the reader. The size, type, and location of all inserts, anchors, and other cast-in items should be defined clearly, and their specific function described. These items should be keyed to a bill of materials shown at the top of the drawing so that a method of checking is readily available to the production staff.

If the panels are cast face down, handling inserts, connection inserts, and cast-in hardware will normally be exposed to view and should be drawn accordingly. Cast-in hardware is shown in as many views as is necessary to ensure proper placement during casting. (See figure 2-71.)

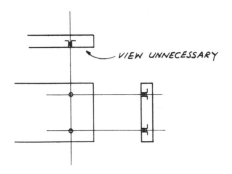

VIEW UNNECESSARY

Figure 2-71

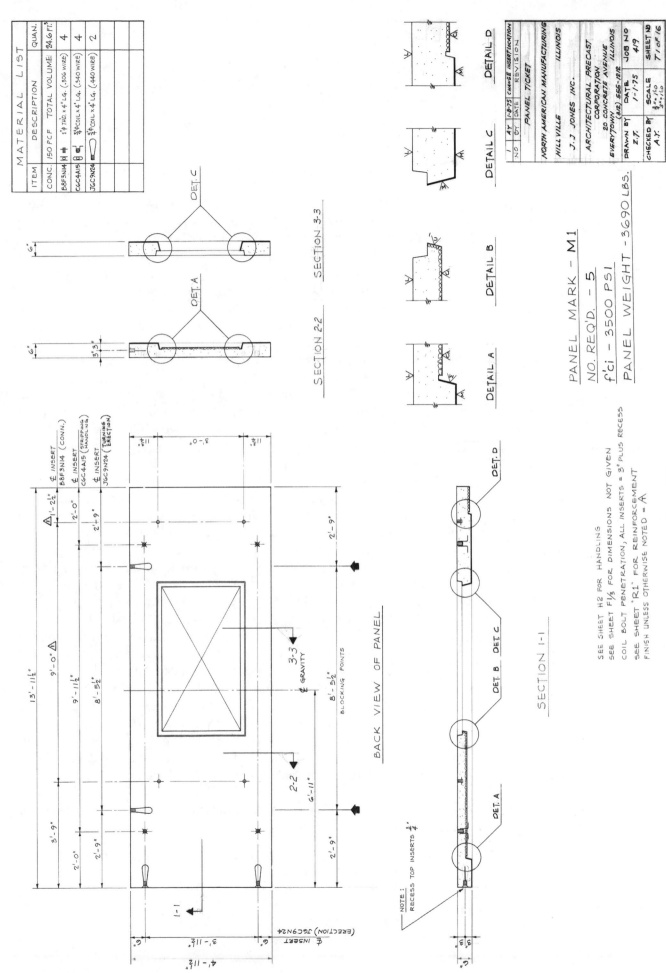

Figure 2-70

Cast-in hardware includes:

a. Inserts (note bolt penetration)
(See Appendix L, page 179.)

d. Reglets

b. Plates

e. Lifting Loops

c. Angles

f. Electrical Boxes and Fixtures

g. Pipe Sleeves

Material that is cast in panels, but supplied by others, should be so noted by giving the supplier's name and material catalog number. All window openings and blockouts should be shown in section with edge requirements detailed to a larger scale, or referenced to the shape drawings. To ensure clarity, an *X* should be lightly drawn to make the opening stand out.

If the center of gravity for geometrically complex panel is not at the panel centerline, it should be indicated and dimensioned for all axes necessary to facilitate stripping and shop handling. (See figure 2-72.) Appendix N shows method of computation for a simple panel.

Blocking points should be included on the panel ticket to ensure proper methods of storage and shipping. This information will also decrease the possibility of damage while panels are stored at the precast plant. If a shortened view is necessary, use break lines and note major dimensions not drawn to scale. (See figure 2-73.)

Enlarged details should be employed when necessary to clarify congested areas or conditions too small to show at normal scale. (See figure 2-74.) Show and detail all drips, (see figure 2-75) false joints, and reveals not already detailed on the form drawing.

Edge requirements around openings should be shown after considering the following items: (a.) shape (consider draft), (see figure 2-65), (b.) finish, (c.) inserts, (d.) reglets, (e.) nailers.

Stripping strength of concrete and cutting and patching to be done in the yard must be noted. Locate and specify shop applied material. Handling instructions for shop, if required, should show location and size of bracing (strongback), inserts, and bracing members. Note when bracing is to be installed and removed.

Define and locate the extent of each type of finish. Be sure to define all surfaces, including back and edges. Do not assume that any finish requirements are "understood" by shop personnel. (See Appendix F.) Show location for panel mark placement, if necessary, to avoid confusion in erection. (See figure 2-22.)

Reference. Make written reference to the following items for necessary information:

a. Shape drawing for panel shape. (Not necessary if panel ticket and shape drawing are the same.)
b. Handling sheet for handling procedures.
c. Reinforcement ticket for cage details.
d. Hardware details.

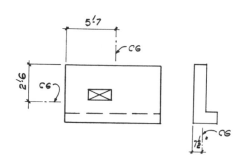

Figure 2-72

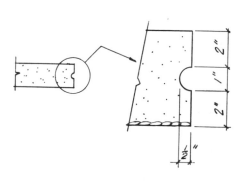

Figure 2-73

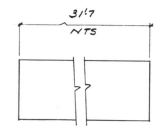

Figure 2-74

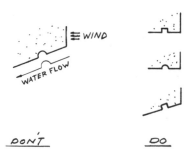

Figure 2-75

General Notations. The following pages will provide the necessary symbols needed to prepare panel tickets:

Dimensioning. Overall panel dimensions must be listed as the outermost dimensions on the panel drawing. They are followed by blockouts, openings, and cast-in items. Blockouts and openings in the panel are dimensioned to their centerline unless they occur at an edge. (See page 178.) The following items are always dimensioned to their respective centerlines:

a. Threaded and Coil Inserts
b. Slotted or Wedge Inserts (See Appendix M, page 180.)
c. Dovetail Slots and Reglets
d. Plates and Angles
e. Pipe Sleeves
f. Lifting Loops
g. Reveals

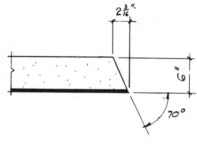

Dimensions requiring extreme accuracy should be noted as critical. Tolerances, if known, should be indicated. If a panel edge is sloped or formed to a certain angle, it should be dimensioned and a degree designation given for secondary reference. (See figure 2-76.)

A final check must be made following completion of dimensioning to make certain that all intermediate dimension strings add up to the overall dimension. All figures must agree with those shown on approved or corrected erection drawings.

Figure 2-76

Cast-In-Items. All items cast into a precast concrete panel must be affixed to the formwork before placement of concrete. This ensures that the concrete completely surrounds the anchors, and aggregates are uniformly distributed. If cast-in items are inserted after the panel is cast, anchors push the aggregates aside, leaving voids in the concrete. This prevents proper consolidation and results in a cast-in item which will not behave as designed.

If inserts are to be recessed, the depth must be noted. (See figure 2-77.)

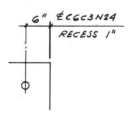

Figure 2-77

In order for the insert to perform to its fullest capability, and to ensure that rust stains do not occur on the exposed surface of the panel, sufficient concrete covering must be maintained. The function of the insert should be noted behind the designation. (See figure 2-78.)

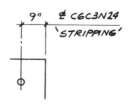

Figure 2-78

Inserts are most commonly utilized for:

a. Stripping
b. Turning
c. Erection
d. Handling

See Appendices P & Q for common handling points and insert loading combinations.

e. Bracing
f. Attachment of Architectural Hardware

Cast-in items should be labeled using the marking system shown on page 50.

Material List. List item, description, and quantity per unit for all material required to produce the panel. Do not include formwork or reinforcing cage. Loose rebar, however, should be listed as it is installed during the casting operation.

HINTS

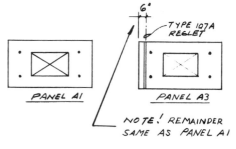

Figure 2-79

a. If the panels are all of different size and type, draw each panel separately, *one panel per sheet*! If these panels have some similarities, the sepia method should be encouraged.

b. Make sepias of the first panel and change dimensions for similar units, if possible.

c. Investigate the possibility of adding extra unused inserts or hardware items to obtain duplication of panels.

d. Draw only one unit per sheet of paper. If more than one panel must appear on a sheet, make certain the minor differences can be readily noticed by using shop note in bold letters (see figure 2-79.)

e. A schedule (tabular) method also may be used to indicate changes in panel features. If a group of panels varies only in length, width, or detail location, prepare one drawing and schedule the different dimensions. (See figure 2-80.) It is necessary to carefully check for interference since items are not drawn to scale.

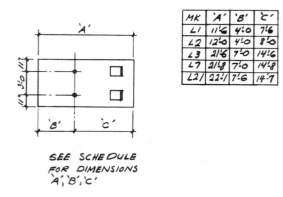

SEE SCHEDULE
FOR DIMENSIONS
'A', 'B', 'C'

Figure 2-80

f. A third method of presentation is by means of schematics. In this method, a typical panel including all details and reinforcing pertaining to that panel is drawn to a scale large enough to show most details. (Other details are shown to a larger scale.) On the same sheet, panels are drawn in a schematic form indicating only variations from the typical panel.(See figure 2-81.) If there is not enough room on one sheet for all variations use more sheets. It is not recommended that the typical panel be shown on other sheets, since it is easy to forget to include revisions on these sheets.

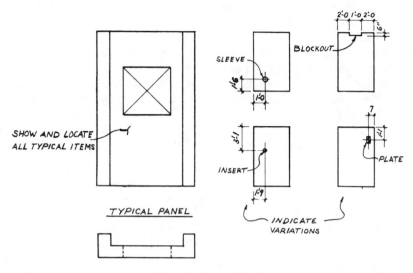

Figure 2-81

g. Indicate where inexpensive back-up mix may be used. It is helpful if there is a monolithic shape. (See figure 2-82.)

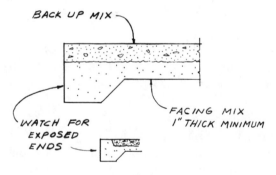

Figure 2-82

h. Note if a panel can be cast in a form previously used on the same project.

i. For production use, do not make reference to details from the erection portion of a drawing.

j. No panel may have more than one mark number.

k. If large blockouts create weakened sections, add reinforcement or inserts for a strong-back to strengthen the panel during handling. Consult with the Precast Engineer before making this decision.

l. Check for additional rebar required around erection and anchorage inserts. Consult Precast Engineer.

m. When using lifting inserts in pairs, check to ensure that center to center spacing conforms to lifting apparatus (12 inch center to center for 1 inch diameter and 15 inch center to center for 1¼ inch diameter bolts usually work well).

n. Ascertain that cast-in hardware does not interfere with the placement of reinforcement.

o. Specify smooth finish wherever connection hardware will be in contact with precast surface.

p. Avoid reductions in area of concrete. (See figure 2-37.) Consult with Precast Engineer if this occurs.

q. It is preferable not to use the same inserts for handling and connections. Handling—coil; connection machine thread. (See Appendix L for recommended bolt penetrations.)

r. Round off cubic footage of concrete to nearest ½ cube.

s. When using numbers to identify panel quantities, spell out the number (one) rather that using the numeral (1).

t. Do not call out the same dimension more than once; dimension each item in one view only. This also applies to inserts, hardware, and reinforcements.

u. For quick calculation, standard weight concrete including reinforcement and hardware weighs 160 lbs. per cubic foot.

Mark	Size	Scale
R	12 x 24 24 x 36	¾, 1

Reinforcing Ticket   Separate reinforcing or cage drawings have become the rule as architectural precast concrete panels have taken on larger and more complicated shapes. A pre-assembled reinforcement cage must mate perfectly with the form in order to prevent production delays. To eliminate interference, layouts and cross checking of integrated materials has become extremely important. (See figure 2-83.)

The reinforcement cage can be shown on the panel ticket if the panel is a simple flat with a cage consisting only of a sheet of welded wire fabric or a few straight bars. A cross section is all that is required.

Planning. If the reinforcing ticket is properly planned, a single reinforcing cage may be employed in the production of similar panels. Determine if a master sepia can be utilized (see page 9).

Allow adequate space for horizontal and vertical sections and blown up details, as well as a material list. (See Appendix V for required information.) Check general notes for grade and finish of steel required. Prepare reinforcing drawings in sufficient time to eliminate any delays in casting.

Presentation. A typical reinforcing ticket layout is provided in figure 2-83. Allow ample space for notations, dimensions, and material lists. Indicate reinforcement cage in the same position as the panel is drawn on the shop ticket. Show dotted outline of panel.

Required Information. The mark number of the reinforcing cage should be keyed to that of the panel in which it is used, except when several panels require the same cage. This number and the quantity of cages required should appear in large print.

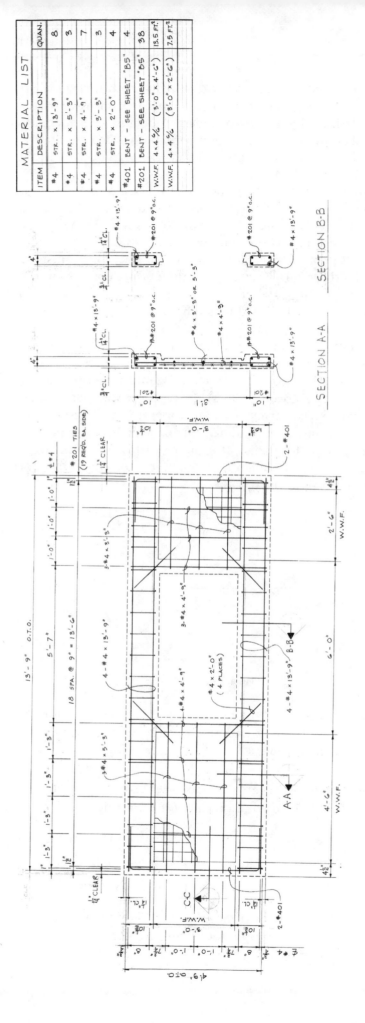

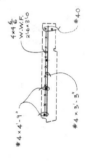
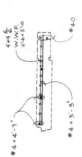

Figure 2-83

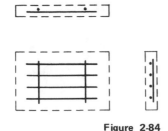

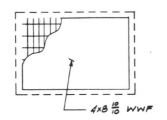

Figure 2-84

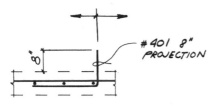

4x8 $\frac{10}{10}$ WWF

Figure 2-85

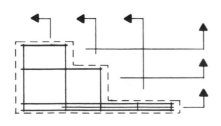

#401 8" PROJECTION

8"

Figure 2-86

Size, dimension, and locate all rebar required to form cage, and list instructions regarding welding, if employed. Congested areas should be clarified by means of enlarged details. Cut adequate sections so that all bent bars, straight bars, and welded wire fabric can be sized and located in plan as well as in section to avoid confusion (see figure 2-84).

Note grade of steel and finish of all rebar (see Appendix T for rebar information). Refer to bar lists for details. When using unsymmetrical welded wire fabric, indicate graphically or dimensionally which direction wires run. (See figure 2-85.)

Show all items other than reinforcing steel that must be pre-affixed to the reinforcing cage. Include a material list which shows item, description, and quantity for all bars (bent or straight) and the size and type of welded wire fabric.

Dimensioning. Overall dimensions of cages should be shown three ways. Indicate clearance from extreme edge of cage to face of precast in each direction. (1 inch min., 1½ inch min. for exposed aggregate concrete.) Locate each rebar (except edge bars) to its centerline and extend strings of dimensions to total the overall dimension. Note any critical dimensions, and indicate projecting length of any steel that penetrates the precast panel. (See figure 2-86.) Draw and define table of rise and radii for curved bars of large radius as shown in figure 2-87.

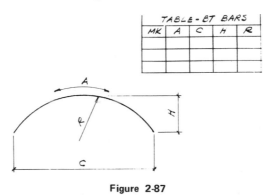

Figure 2-87

Any bars projecting through the form should be dimensioned and located. When locating tied reinforcing follow dimensioning practices set out in Appendix T.

General Notations. The following pages show the exact notations and symbols to be used in preparing the drawings:

Abbreviations . . . . Appendix D, page 150

Drawing Symbols . . . . Appendix F, page 163

Welding Symbols . . . . Appendix F, page 171

Sections. Cut a section through each change of cross section both horizontally and vertically. (See figure 2-88.)

Figure 2-88

References. The bend details may be obtained from the bent bar list (see Appendix T.)

HINTS

a. Depending upon the amount of room available, bent bar details can be shown either on the cage drawing or on separate cut sheets.

b. Use notations which cannot be misinterpreted such as "each way" rather than "both ways"; "#5 @ 12 each face, staggered" rather than "#5 @ 12 staggered".

c. Where the form prohibits projecting reinforcement, obtain approval to substitute cast-in inserts and threaded rods or coil rods. (See figure 2-89.)

d. Check, by making full size layouts of any potential problem areas: several bars in different planes meeting; the radius of a bent reinforcing bar may be critical concerning clearance; or bars passing beyond hardware with anchors. (See figure 2-90.)

e. Projecting bars with exposed bends should be shown straight and bent following stripping. (See figure 2-91.)

f. Cold-bending of reinforcement and shearing and punching of plates shall not be permitted if the finished item is to be galvanized, unless it has been normalized between cold-working and galvanizing. The purpose of this rule is to prevent the acceleration of natural strain aging by the heating involved in pickling and hot-dipped galvanizing on steel which has been severely deformed by tight bending, punching, and shearing. There is also the likelihood of hydrogen embrittlement of the steel being caused by acid pickling due to the entry of hydrogen into micro-cracks at bends and sheared surfaces.

g. Never weld a rebar without approval of the Precast Engineer.

h. Note special practices to be followed when welding high-tensile steel, such as preheating and use of low hydrogen electrodes.

i. Use high-tensile steel only when specified in contract drawings.

j. Detail bars to three inch increments of length if possible. This is helpful for takeoff purposes. Allow clearance to increase slightly. (See figure 2-92.)

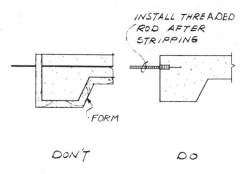

INSTALL THREADED ROD AFTER STRIPPING

DON'T   DO

Figure 2-89

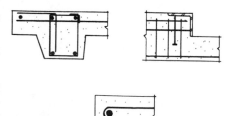

Figure 2-90

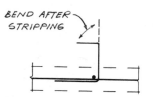

BEND AFTER STRIPPING

Figure 2-91

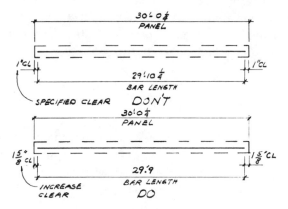

Figure 2-92

k. Do not place parallel bars closer together than the size of the maximum aggregate used. (See Appendix H, page 175.)

l. Place rebar 1 inch clear of large openings and provide corner bars as required in accordance with the Precast Engineer's calculations.

m. To provide the reinforcement assembly with adequate handling stability, it may be necessary to add extra bars.

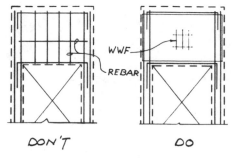

Figure 2-93

n. When reinforcing flat areas supplement WWF (welded wire fabric) for rebar grid. (See figure 2-93.)

o. Graphically displace bars from actual plane, if clarity is increased. (See figure 2-94.)

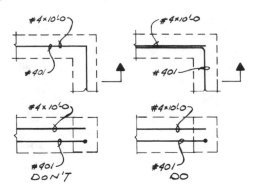

Figure 2-94

p. If possible, reinforcing cages should be supported from the back of the panel, because spacers of any kind are likely to mar the finished surface of the panel. For smooth cast facing, stainless steel chairs are tolerable, if specifically permitted in the specifications. In general, metal chairs with or without coating should not be used in the finished face. Some precasters have found plastic chairs to perform satisfactorily.

q. Use welded wire fabric wherever possible (see Appendix S for common styles).

Mark	Size	Scale
B	8½ x 11 †	none

† 24 x 36 quadranted

Bar List   A bar list is a grouping of all reinforcing bars and welded wire fabric—straight or bent—required for a given project. Only the steel used in the fabrication of precast units is included in the list. Rebar which is supplied loose at the job site is listed as erection material.

The information that follows is intended as a guide in the preparation of bar lists. (See figure 2-95.)

Planning. Establish a master schedule sheet similar to the illustration in figure 2-95. Make several copies of this master schedule. Group all bars and welded wire fabrics by stock size*, and indicate grade of steel and whether bent or straight. Group together bars with similar types of bends.

*For rebar use ASTM (*American Society for Testing and Materials*) bar size designation, see page 194 and for welded wire fabric use WRI (*Wire Reinforcement Institute, Inc.*) style designation. See page 189.

			Length												
Mark	Req'd	Size	Ft	In	Shape	Grade	Finish	A	B	C	D	E	F	G	Shape Key
	40	#4	13	9	1	40	BLACK	13'-9"							
	15	#4	5	3	1	40	"	5'-3"							
	35	#4	4	9	1	40	"	4'-9"							
	15	#4	3	3	1	40	"	3'-3"							
	20	#4	2	0	1	40	"	2'-0"							
#20l	190	¼"	2	11½	2	40	"	4	10	4	10				
#40l	20	#4"	7	6½	3	40	"	1'-6"	4'-7"	1'-6"					
	5	4x4 ⁶⁄₆	4	6	1	185	GALV	3'-0"							
	5	4x4 ⁶⁄₆	2	6	1	185	GALV	3'-0"							

Bar List — FOR RI CAGE (PARTIAL)

By 2T
Job N.AM.
Date _____
No 419
Checked AY
Sheet B5 OF 21

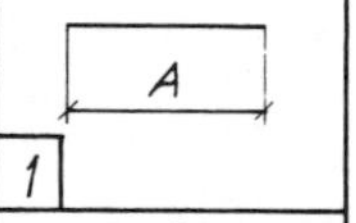

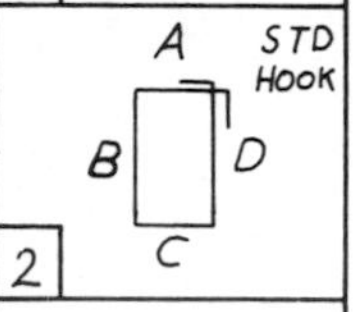

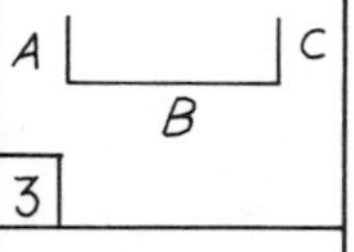

Figure 2-95

Basic Presentation. Use the following separate groupings:

> Straight Bars
>
> Bent Bars
>
> Sheet of Welded Wire Fabric
>
> Bent Sheet of Welded Wire Fabric

The straight bars are grouped according to size with the largest size first. Those of the same size are listed in the order of their length with the longest bar first. This graduation in length saves time and materials in the shearing (cutting) operation. Dimension bends or make reference to bending diagram.

Required Information:

Straight Bars: size, length, grade, finish, quantity.

Bent Bars: size, length of bar to be bent, grade, finish, quantity, angle of all bends, all out-to-out dimensions after bending.

Flat Sheets of Welded Wire Fabric: size, length, width, finish, quantity.

Bent Sheets of Welded Wire Fabric: size, finish, quantity, out-to-out size of sheet to be bent, angles of all bends, out-to-out dimensions after bending.

Clearly note any change in the grade of steel. Note if galvanizing is required. Indicate panel marks, sheet number, or area of project for which material is intended. State whether list is complete or partial.

Marking. Mark all material for easy reference.

Straight Bar . . . . #3 x 11'6"

Bent Bar . . . . #301 (#3 rebar, type 1)

Flat WWF . . . . 4 x 4 4/4, 3'0" x 6'0"

Bent WWF . . . . WWF–A (Bent WWF–Type A)

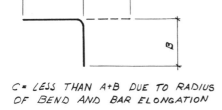

C = LESS THAN A+B DUE TO RADIUS
OF BEND AND BAR ELONGATION

Figure 2-96

HINTS

a. When figuring pre-bent length of bars, make allowance for gain encountered during bending. (See figure 2-96.)

b. Follow ACI (*American Concrete Institute*) bending recommendation at all times. To avoid creating excessive stresses during bending, bars must not be bent too sharply. Controls are established by specifying the minimum inside radius or inside diameter of bend which can be made for each size of bar. (See Appendix T for rebar bending information.)

c. Alter clearances to change straight bar lengths in 3-inch increments only. (See figure 2-92.)

Mark	Size	Scale
M	8 1/2x11†	none

† 24x36 quadranted

When all of the production drawings are completed, a production bill of materials should be prepared. This is a matter of adding together all of the items on the individual production drawings— the panel and reinforcing tickets—and summarizing on a material list.

Production Material List

The information that follows provides a guideline for the preparation of the production material list. (See figure 2-97.)

PRODUCTION MATERIAL LIST *(PARTIAL)*		DATE _____ — _____ BY _____ VL _____ CHECKED AY		JOB ___ N.A.M ___ NO ___ 419 ___ SHEET M9 OF 16	
ITEM	QUANTITY	DESCRIPTION	SUPPLIER	REMARKS	DATE ORDERED
CONC*	180 CU FT	MIX # 241		SHOP MIX	
# 4	10	13'-9 LG	SMITH FAB		
# 401	6	SEE SHT B-3	SMITH FAB		
WWF	23.5 SQ FT	4 x 4 6/6 2'-9 x 8'-6		1 PC ONLY	
INSERT	130	F6 612 A16	FLASH CO	GALVANIZED	
REGLET	36 LIN FT	TYPE 107A	ACE SUPPLY		
L-5-5	8	SEE SHT S5	SMITH FAB		
INSUL	128 SQ FT	1" X 2'-0 X 8'-0 POLYSTYRENE		8 PCS	

*Can be broken down (coarse aggregate, sand, cement, etc.)

Figure 2-97

Planning. Establish a master list similar to the example in figure 2-97, and make required copies from the master. Divide material into similar groupings for logical presentation.

Basic Presentation. List quantity, description, supplier (particularly when a nonstandard item is ordered) and date required for all materials. Good estimates of quantity requirements will allow the purchaser to take advantage of job lot prices.

Required Information. The following should be listed:

1. Concrete—total cubic feet of each mix.
2. Straight and Bent Bars—total lineal feet and pounds of each bar size. (Make reference to bar list for breakdown.)
3. Flat and Bent Welded Wire Fabric—total square footage of each welded wire fabric size. (Make reference to bar list for breakdown.)
4. Inserts—total number of each type.
5. Reglets and Slots—total lineal feet of each type.
6. Strand—total lineal feet of each size.
7. Hardware—total number of each mark.
8. Insulation—total square feet of each thickness.

HINTS

a. To avoid delays in production, issue production material list as early as possible.
b. To obtain total quantities, use a take-off sheet similar to that in figure 2-98.
c. Overorder on critical materials to compensate for waste.

PANEL REQ'D	INSERT H6F3N7	#401	L-S-5
B1 / 6	4 / 24	20 / 120	
B2 / 2	4 / 8	20 / 40	
B3 / 2	6 / 12	20 / 40	
B4 / 1	2 / 2	10 / 10	
TOTAL	46	210	

Figure 2-98

Mark	Size	Scale
H	8-1/2x11†	none

†24 x 36 quadranted

Erection Material List When the precast firm is erecting as well as producing the units, an erection bill of materials will be required. The erection material list is a summary sheet of all materials employed in the installation of precast units. Elements of the list are loose hardware, sealant, and all other materials which are shipped separately to the job site.

The information that follows is intended as a guide in the preparation of the erection material list. (See figure 2-99.)

Planning. Devise a master list similar to the example in figure 2-99 and make required copies from the master. All materials should be divided into like groupings for logical assembly. When handling sheets for the precast units are drawn, these sheets may be listed in the erection bill of materials to tell the erecting crew that they are available.

Presentation. List quantity, description, supplier, and date required for all materials.

Required Information. The following materials should be listed:
1. Hardware—total number of each mark
2. Bolts and Washers—total number, size, and diameter of each type.

3. Grout—total cubic feet of each mix.
4. Sealant—total number of gallons of each type.
5. Paint—total number of gallons of each type.
6. Sealers—total number of gallons of each type.
7. Joint back-up—total lineal feet of each type for caulking.

HINTS

a. Issue erection material list as early as possible to prevent delays at job site.
b. Use a standard take-off sheet similar to figure 2-99 to obtain total quantities.

Checking

It is mandatory that all erection drawings be checked prior to submittal, and that all production drawings be thoroughly checked before being sent to the shop. When these drawings have been completed to the satisfaction of the project draftsman, a check set is printed. Each print is stamped "Check Print", and dated. The prints are then forwarded to the checker for approval.

The checker will indicate all required corrections in red, and cross out all other correct information in yellow. This procedure is followed to ensure that everything has been reviewed. The project draftsman will back-check and verify all required corrections when the prints are returned by the checker.

Since the project draftsman is responsible for the accuracy of the drawings, no changes will be made unless he is in complete agreement. When the red corrections are made on the tracing, they are encircled in green on the check print.

Once all corrections have been made, the check prints and original tracings are given to the checker. He will complete all necessary back-checking and initial the original tracings.

Prior to issuing final approval and initialing the original tracing, any print issued, other than for approval, will be stamped "Preliminary".

It is advisable to check pessimistically by attempting to derive the answers, rather than justifying the ones shown.

HINTS

a. Spot check sections, elevations, and details for one piece.
b. Check take-off sheet for piece total comparison.

Revisions

Revisions may be required at any time during the course of a project. They must be dealt with immediately in order to keep expensive corrective measures to a minimum.

Listed on page 75 are a number of typical procedures used in dealing with revisions.

Erection Material List

(PARTIAL) #10

Date —
By VL
Checked AY
Job N.A.M.
No 4/9
Sheet H2-4

Item	Quantity	Description	Supplier	Remarks	Date Ordered
Q-E-7	8	SEE SHT S17	SMITH FAB	GALVANIZED	
BOLT	130	3/4"φ X 1 3/4 MACH THRD	ACE SUPPLY	A 307	
WASHER	130	1/8" THICK FOR 3/4 φ BOLT	ACE SUPPLY	A 36	
GROUT	72 GAL	TYPE R11-A	CONST SUPPLY	PRE-MIXED	
SEALANT	130 GAL	TYPE LE990	CONST SUPPLY		

Figure 2-99

1. Place "hold" on any work in progress which may be affected.

2. Retrieve affected prints.

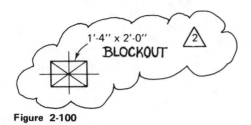

Figure 2-100

3. If the change is minor, make correction, encircle it on the reverse side of the sheet, revise the date, and re-issue to the shop. (See figure 2-100.) The circle is erased following any later revision without destroying the drawing.

4. If additional costs are involved, check estimate and contract documents to determine responsibility.

5. Written approval should be obtained from the Contractor for any change in the total cost of the contract.

6. Obtain approval of Architect/Engineer of Record for any change or alteration to design intent.

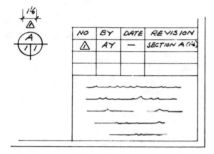

Figure 2-101

7. The Architect must approve any change involving appearance or function of the structure.

8. Any change in drawings, following initial submittal, must be clearly marked with a revision indicator. In standard title block, show revision indicators, date of revisions, and reason for revisions (be brief but encompassing). Refer to figure 2-101 for example of revision recorded on drawing.

9. Indicate on drawings any changes which have been made other than those requested by Architect/Engineer.

10. To facilitate location, mark original tracings with red revision indicators.

11. Use red ink for rubber stamps indicating revision.

12. If a second revision must be made erase all indicators of the first revision. Only one set of revision symbols should appear on a given sheet at any time.

A. CHECK LIST FOR STARTING JOB

(　　) Assign or determine job number
(　　) Set up job file
(　　) Determine lead time
(　　) Check building code
(　　) Read specifications
(　　) Update contract drawings per addenda
(　　) Study contract drawings
(　　) Read bid take-off
(　　) Locate all units listed on take-off
(　　) Check precast support conditions
(　　) Check for production, shipping, and erection problems
(　　) Record and file all decisions or revisions made
(　　) Determine drawing size requirements
(　　) Determine drawing procedure: elevation or plan
(　　) Draw cover sheet

B. COVER SHEET CHECK LIST

Include:

(　　) Company logo
(　　) Engineering Department phone number
(　　) Project draftsman's name
(　　) Project name
(　　) Precaster's job number
(　　) Telephone directory
(　　) Site map
(　　) Shop notes
(　　) Quality control notes
(　　) Erection notes
(　　) Miscellaneous notes
(　　) Contract document list and latest revisions
(　　) Erection drawing list
(　　) Standard abbreviations
(　　) Drawing symbols
(　　) Graphic symbols

C. KEY PLAN CHECK LIST

(　　) Show grid lines
(　　) Show north arrow
(　　) Show existing roads
(　　) Indicate all buildings within projects
(　　) Show existing buildings
(　　) Show variations in vertical height
(　　) Label plan

( ) Indicate scale used
( ) Show locations of adjacent utilities and/or structures
( ) Give erection sequence
( ) Locate underground items

D. SCHEDULE CHECK LIST

Include:

( ) Unit mark
( ) Number required
( ) Maximum length
( ) Unit weight
( ) Maximum width
( ) Maximum depth
( ) Section type if required
( ) Connection type (top, middle "if required", bottom)
( ) Finish (back, front, ends, top, bottom)
( ) Ticket number
( ) Cage mark
( ) Handling detail or code
( ) Sheet found
( ) Remarks

E. ELEVATION/ERECTION PLAN CHECK LIST

( ) Indicate building north
( ) Draw elevations in one plane only
( ) Plot plan—indicate portion shown
( ) Show each panel at least once
( ) Label each plan or elevation
( ) Give parallel grids
( ) Use enlarged details to avoid congestion
( ) No opposite hands or lefts and rights
( ) Show all pertinent grids, floor lines, and steel elevations
( ) Define structure and variations
( ) Identify adjacent structures or materials
( ) Note scale used
( ) Show and size all openings, blockouts, returns
( ) Locate connections typically
( ) Cast-in-place hardware list
( ) Handling schedule
( ) Use 3-key system for sections
( ) Show height and width of each type panel
( ) Size joints
( ) Tie panel edges to grids and floor lines
( ) Give overall dimensions and grid to grid
( ) Mark each unit

(   ) Refer to shape drawing and anchor plan if necessary
(   ) Cross reference match lined areas
(   ) Request verification at questionable areas
(   ) Call attention to changes for verification
(   ) Give elevations (top of steel, top of precast, finished floor)
(   ) Give erection sequence if required

F. SECTION CHECK LIST

(   ) Note scale used
(   ) Define all finish requirements
(   ) Show and call out size of structural support
(   ) Show architectural surroundings if pertinent
(   ) Give ± elevations corresponding to elevation or erection
        plans
(   ) Show and locate grid lines each way
(   ) Identify and locate all connection hardware (inserts, shims, bolts,
        loose hardware)
(   ) Dimension joints and specify joint material
(   ) Specify, size, and locate reglets, nailers, and reveals
(   ) Locate and give size and length of welds
(   ) Refer to anchor plan, shape drawing, elevations, and similar sections
(   ) Use 3-key reference system
(   ) Size, locate, and give function for all openings, blockouts, etc.
(   ) Enlarge details for clarity
(   ) Dimension clearances, hardware, joints, finish limitations, panel
        configuration, and tie panel to ± elevations
(   ) Check dimension totals
(   ) Note critical dimensions
(   ) Request verification if required
(   ) Flag intentional changes
(   ) Use finish separation strip if required
(   ) Show erection sequence if required
(   ) Specify recessed anchors as required
(   ) Note erection restriction and/or special requirements
(   ) Size and locate reinforcement, if required for approval
(   ) Note any material or item not furnished as "by others"
(   ) Show and locate all connections (three dimensionally) and give
        function
(   ) Show flashing type and location (watch for returns)
(   ) Ensure that reasonable initial and long term tolerance requirements
        can be met
(   ) Carry all dimension lines full length or width
(   ) Ensure adequate sectioning:
              each cross section   joints
              each connection condition   corners (interior, exterior)
              openings   cast-in items

G. HANDLING DETAIL CHECK LIST

Illustrate:

(   )   Stripping
(   )   Turning
(   )   Temporary blocking
(   )   Storage
(   )   Shipping
(   )   Erection
(   )   Shoring
(   )   List panel mark(s) covered
(   )   Locate inserts used
(   )   Show stacking orientation
(   )   Indicate crane lines and angles of lines at erection
(   )   Spreader beam if required
(   )   Coil bolt lengths
(   )   Dunnage and/or padding
(   )   Trailer type

H. ANCHOR PLAN CHECK LIST

(   )   Call out size and locate structural components used for attachments
(   )   Show all pertinent grid lines
(   )   Note scale used
(   )   Label plan and note whether partial or complete
(   )   Indicate building north
(   )   Define depressions or raised portions
(   )   Indicate size and type of all cast-in hardware
(   )   Section each different type and application of hardware
(   )   Show rebar which must clear anchors and request relocation if required
(   )   Note items shown but not furnished as "By Others"
(   )   Locate anchors to centerline
(   )   Ensure that anchors are dimensioned to an accessible point
(   )   Note critical dimensions
(   )   Flag intentional changes
(   )   Ensure reasonable initial and long-term tolerances
(   )   Set back pre-welded connections
(   )   Provide plot plan and note portion shown
(   )   Show embedded and projecting length for anchor bolts
(   )   Cast-in material list

I. HARDWARE DETAIL LIST

(   )   Use proper nomenclature
(   )   Give material size
(   )   Give hole or slot size

(   )   List anchor type and length
(   )   Give welding instructions
(   )   Note insert types
(   )   Specify finish
(   )   Specify material type
(   )   Give special instructions
(   )   Give unit weight
(   )   Follow proper hole locating procedures
(   )   Nail holes
(   )   Bleed holes

J. SHAPE DRAWING CHECK LIST

(   )   Provide reasonable drafts
(   )   Show view of each formed face
(   )   Indicate panel(s) and quantity
(   )   Indicate finish requirements (all surfaces)
(   )   List tolerances
(   )   Section each cross-sectional variation
(   )   Carry dimension lines full length or width except for drafts
(   )   Give strand locations if prestressed
(   )   Locate edge inserts
(   )   Show or provide radii or chamfer at exposed edges
(   )   Note intentional changes
(   )   Request verification of questionable details or dimensions
(   )   Set drawing up for later conversion to panel ticket if practical
(   )   Ensure that largest unit can be obtained from form or mold
(   )   Make provision for modification to other similar panels
(   )   Draw to largest scale possible

K. PANEL TICKET CHECK LIST

(   )   Draw one type panel per sheet unless using schematic or tabular method
(   )   Show view as cast
(   )   Size, locate, and define hardware:

> handling inserts or loops
> stripping
> turning
> erection
> bracing
> connection
> plates, angles, etc.
> reglets
> sleeves
> mechanical fixtures

(   )   Note material to be supplied by others
(   )   Size, locate and define all openings and/or blockouts

(   ) Give centers of gravity (3 ways)
(   ) Show and locate blocking points for all handling procedures
(   ) Use blown-up details to avoid confusion
(   ) Size and locate reglets, drips, false joints, reveals, returns, etc.
(   ) List panel mark, number required, concrete strength at stripping, and unit weight
(   ) Refer to shape drawing, reinforcing ticket, handling details, bar list, and hardware details
(   ) Show or note separately any operation or installation which is to take place after stripping
(   ) Specify and locate position of strongback if required
(   ) Give overall dimension (3 ways)
(   ) Note important dimensions as "critical"
(   ) List tolerances
(   ) Check dimension totals
(   ) Make form attachment provisions for all cast-in material
(   ) Specify recessed inserts if required
(   ) Check concrete cover for all cast-in items
(   ) Note and dimension back-up concrete if used
(   ) Add unused inserts if standardization can economically be achieved
(   ) Avoid reduction in concrete section
(   ) Provide loose rebar around or through inserts
(   ) Use coil inserts for handling
(   ) Use machine threaded inserts for connections
(   ) Round concrete volume(s) off to nearest half of a cubic foot
(   ) Include material list
(   ) Define and locate all finish requirements
(   ) Size and locate insulation if required
(   ) Specify placement of panel mark if used for orientation
(   ) Check connection insert locations against anchor plan

L. REINFORCING CHECK LIST

(   ) Show dotted panel outline
(   ) Orient same as panel production ticket
(   ) Show mark and quantity
(   ) Give welding instructions including electrode if employed
(   ) Show each bar in section twice (once horizontally and once vertically), identify and locate
(   ) Note grade and finish of materials
(   ) Indicate direction for unsymmetrical WWF
(   ) Refer to bar list for bends
(   ) Give overall dimensions (3 ways)
(   ) Show all clearances to P/C (precast concrete) edges
(   ) Note critical dimensions as such
(   ) Indicate projecting length and location of any projecting steel
(   ) Cut sections at each cross-sectional variation
(   ) Study complicated intersections via full size layout
(   ) Detail bars to three inch increments of lengths
(   ) Locate bars or ties to centerline

(   ) Include material list
(   ) Ensure that cast-in hardware will clear all reinforcement
(   ) Consider aggregate size when placing parallel bars
(   ) Add bars as required to provide handling stability to the cage as a unit
(   ) Show threaded-on hardware

M. BAR LIST CHECK LIST

(   ) Detail all bends
(   ) Give size, shape, quantity, material grade, and finish for:
     straight bars
     bent bars
     sheets of WWF
     bent WWF
(   ) State whether partial or complete list
(   ) Note areas of project covered
(   ) Allow for gain due to bending
(   ) Change straight bar lengths in 3-inch increments

N. PRODUCTION AND ERECTION MATERIAL LIST CHECK LIST

(   ) List number of pieces of bent rebar, straight rebar, fabricated hardware, bolts, inserts, etc.
(   ) List lineal feet of cable, reglets, nailers, etc.
(   ) List square feet of insulation, mesh, etc.
(   ) List cubic feet of concrete, grout, etc.
(   ) List gallons of sealant, paint, sealer, etc.
(   ) Designate supplier if possible
(   ) Indicate partial or full list
(   ) Note portion of project represented

O. CHECK LIST FOR JOB COMPLETION

(   ) Transmit final prints to all parties
(   ) Sort all work sheets, sketches, etc.
(   ) File all calculations, sketches, and letters
(   ) Roll-up, mark and store all approvals, work sheets, contract drawings, etc.
(   ) Mark and store project specifications and addenda
(   ) Store original drawings
(   ) Double check to ensure that all items are in order and properly filed or stored
(   ) Inform immediate superior of completed status

THE SUBMITTAL PROCESS

Once the precast erection drawings have been completed and checked, they are dated and submitted through the proper channels for approval. All deviations from the contract documents should be clearly noted for approval. The following discussion provides an explanation of the submittal process for erection and shape drawings.

Section III

Only erection and shape drawings are submitted for approval, although the Architect/Engineer may request record prints of production drawings.* No portion of the precast concrete work is permitted to start until the submission has been approved.

Before starting the work, the Contractor submits to the Architect/ Engineer for approval a schedule of drawing submissions and a conference usually is held to review the procedure for handling the drawings.

Submittal Sequence

The period of time between the letting of the contract and the delivery date will determine the sequence of drawing preparation for·submittal. There are two submittal sequences that are practiced throughout the architectural precast concrete industry. The first sequence for transmittal is the most ideal from the Precaster's standpoint. However, the second sequence is the most commonly used method from the standpoint of practicality.

SEQUENCE 1

a. Cover sheet, key plan, elevations, erection plans and sections. In this first step it is not necessary to show connection details because initial drawings should be submitted for shape and dimensional approval only.
b. Shape drawings.
c. Cover sheet, key plan, elevations, erection plans, sections, hardware details, schedule and handling details for final approval.
d. Anchor plans.

Unfortunately, there is seldom enough time to submit erection drawings as shown in Sequence 1. When lead time is short, the project may cost the owner more than it should. When time is short, Sequence 2 must be used.

SEQUENCE 2

a. Anchor plans
b. Shape drawings
c. Key plan, cover sheet, elevations, erection plans, hardware details, handling details, schedules, and sections showing connections for final approval.

By submitting anchor plans and shape drawings for approval first, a number of problems may occur: anchors may be omitted for units not properly defined on the contract documents; anchor locations may be missed; form modifications may be required later because certain requirements were not readily discernible at the time drawings were prepared; or later modifications may require changes in forms. *It is therefore recommended that Sequence 1 be*

*In addition to drawing submittal, calculations pertaining to the loads and movements used in the design of the panels shall be submitted to the Architect/Engineer, if requested.

employed whenever possible and the importance of lead time be emphasized to the Contractor. For large or major projects, the drawings may be submitted for approval in installments. For example, for multi-story buildings erection drawings may be submitted for approval one or more stories at a time.

Method of Submittal   The method of processing drawings is a matter of individual preference. Usually a reproducible transparency (sepia) is accompanied by a form letter of transmittal which is simply filled in to record such data as the identification of the project, contractor's name, the drawings, their source, number of copies, the date sent, and any action required. (See figure 4-9.) On large projects it is necessary to keep a log of such data, and also the date returned and such results of the action as "approved," "approved as corrected," "revise and resubmit" or "not approved."

Prior to releasing an erection drawing for approval, indicate release date. Use space provided in title block if available. This enables a check against the letter of transmittal and revision change dates.

Specifications usually state the number of copies required for approval. The most advantageous method of submitting precast drawings for review and checking is by means of a sepia. The approver makes a nonreproducible print which he thoroughly checks. He then transfers all pertinent information from the print to the sepia. The added information is then encircled with a heavy dark line so as to be readily visible. Prints are then made from the sepia for distribution to all concerned parties. This submittal procedure has three advantages:

1. The time consuming process of transferring information to several sets of prints is eliminated.
2. All persons receiving approval prints receive exactly the same information.
3. Eliminates possible errors of omission through failure to record comments or revisions on all copies of submittal drawings.

From the approver's point of view, making record copies has definite disadvantages. They may increase printing costs, and unapproved changes may be made on the reproducibles which appear on prints made after his approval stamp has been affixed.

Routing   The routing procedure for precast drawings often takes a complex course. In most cases, the submittal procedure is covered by the General Conditions of the project specifications. It is not uncommon for this procedure to cover a two- or three-week period. However, prompt approval of the drawings is essential to meet production schedules and delivery dates. Maintaining the progress schedule is the responsibility of the Contractor.

An understanding of the routing procedure will perhaps allay some of the frustrations encountered while waiting for approval. Drawings should be forwarded to the Architect for his review only when and if they have been checked and approved by the Contractor. When consultants are given responsibility for such approvals, the extent of this responsibility must be agreed upon and understood by all concerned. The following is a typical routing plan during the submittal period:

1. Precaster to Contractor
2. Contractor to Various Trades
3. Trades to Contractor
4. Contractor to Architect
5. Architect to Engineer(s) of Record
6. Engineer to Architect
7. Architect to Contractor
8. Contractor to Precaster
9. Contractor to Various Trades

In certain instances, drawings may also be routed through local building departments.

Approval

Checked drawings are returned to the Precaster with one of the following notations:

1. Approved—No resubmission necessary.
2. Approved as Corrected—No resubmission necessary.
3. Revise and Resubmit.
4. Not Approved.

Most offices have a rubber stamp containing these notations. The stamp may also contain a statement as to the responsibility assumed, as well as that not assumed, by the approver. In some cases, the draftsman should provide a space near the title block on each sheet to accommodate the stamps of the Architect and/or Engineer and the Contractor (check Supplemental General Conditions of specifications). *In order to receive a binding approval, it is necessary that every sheet in the set—not just the first sheet—be stamped by all who approve it.* On approval, these stamps make the drawings an official part of both the contract documents and the specifications.

After approval, any desired changes in panel drawings which are necessary to obtain a proper standard of manufacture should be incorporated only after the structural effect of such changes has been approved by the Engineer of Record.

When the approving agency returns prints with notations for corrections or stamped "not approved", the reasons for such action should be indicated, and all changes must be checked. These changes may affect other work in progress in the drafting room, or work already released for production.

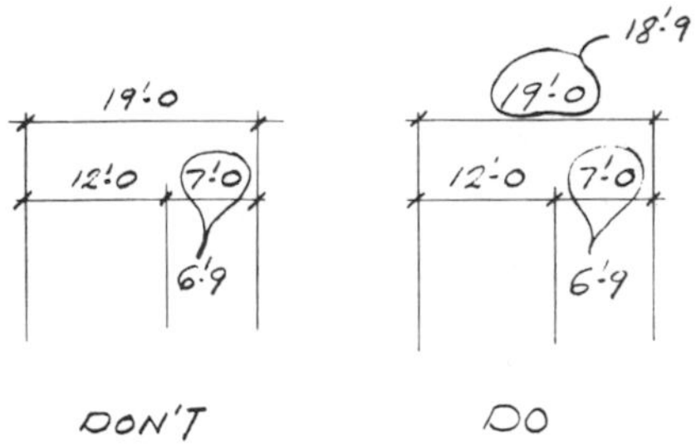

DON'T   DO

Figure 3-1

It is mandatory that the approver follow through with all dimensional changes rather than simply note one change in a strong of dimension. (See figure 3-1.)

If found to be in order the changes should be made promptly, and revised production drawings issued to the shop. If required, new copies of the corrected drawings are then resubmitted for final approval and/or field use.

Too often the definition of approved drawings is not clear. Sometimes the parties approving drawings are approving for concept only. Since the precast drawings are developed from the architectural and structural framing plans, it is not possible nor practical for the Precaster to be responsible for building dimensions—it is the General Contractor's responsibility. However, in preparing the drawings should the Precaster discover any error, inconsistency, or omission in the design drawings, he must refer them at once to the approving agency. Further instructions must be received before proceeding with the affected part of the work. Should the Precaster fail to detect such discrepancies in the design drawings, the responsibility for any extra costs that may result from them still rests with the General Contractor. Because of the involved contractual relationships which often exist (between the Precaster and his customer, other sub-contractors of this customer, and the owner and his Engineers and/or Architect), the Precaster must report in writing any discrepancies on the design drawings, or in the contract specifications, at the earliest possible moment. The draftsman is responsible for reporting these discrepancies to his superior.

An additional aspect to consider relative to precast concrete erection drawing approval is that the Contractor and/or Architect/Engineer may be aware of changes in building dimensions, building frame modifications, or requirements of other trades, and must continually communicate these facts to the Precaster. It is the Contractor's responsibility to coordinate the precast concrete with the other trades.

"Approved" or "Approved as Corrected" drawings should mean that the General Contractor has verified dimensions to be correct and final for the following: overall building dimensions, column centerlines, floor elevations, the locations of mechanical openings, and other items pertinent to the architectural precast concrete. Thus, the approved drawings tell the precast producer that he is now responsible *only* for coordinating the precast product dimensions to the building dimensions. These building dimensions can now be taken as correct and final. The General Contractor must immediately notify the Precaster of any deviations found in dimensions due to plan or construction errors.

The Precaster can proceed with his production drawings based solely on his approved erection drawings. He need make no further changes, and may now issue erection drawings for field use. If shape drawings are submitted separately, approval would allow fabrication of molds and tooling.

Approved

The Precaster may proceed as described above, but not before making the noted corrections, unless the notation or contract documents specifically prohibit such action. He need not resubmit the erection drawings prior to beginning production drawings, but should do so later for final field use and as a record copy.

Approved as Corrected

The Precaster cannot proceed with the production drawings based upon the drawings he submitted. He must first make the required changes, resubmit, and then wait until the second submittal is returned. Specific attention in writing should be directed to revisions on the resubmitted drawings other than the corrections requested on previous submissions.

Revise and Resubmit

Appendix B discusses design and construction responsibilities of the Architect, Engineer of Record, Contractor, and Precaster. Although there are several options which allocate varying degrees of responsibility to the various parties, "Approval" means the following unless otherwise stated:

Responsibility of Approver

> Approval by the *General Contractor* indicates to the Precaster that all as-built building conditions at the time of erection of precast will be in accordance with the approved precast erection drawings and within recognized industry tolerances. It tells the Precaster that the Contractor has completely checked all design building dimensions. These include column centerlines, floor elevations, panel locations, sizes and shapes, and location of connections and finishes. Approval also indicates that all requirements by other trades, such as electrical, mechanical, heating, ventilating, and window suppliers, have been properly interpreted. Unfortunately, some contractors automatically stamp the drawings approved and submit them to the Architect, expecting him to find and work out the details of all variances from the original contract requirements. Hence, the Precaster should be extremely careful to be sure that his design conforms to the plans and specifications in all ways so as to avoid controversy as to whom is responsible for any errors or omissions.

> Where specifications call for approval by *Architect* and/or *Engineer of Record*, they review and approve the drawings only for conformance with the design concept of the project

and with the information given in the contract documents. However, the drawings frequently contain information which is not related to the design concept or information that is relative only to the production process or construction techniques in the field, all of which are outside the scope of the Architect/Engineer's duties and responsibilities. Therefore, they should always word their approval of these submissions in such a way so as not to indicate approval of such data. Approval by the *Architect* also indicates to the Precaster that all details such as size, shape, and finish conform to his aesthetic requirements. The Architect's decisions in matters relating to artistic effect are final if consistent with the intent of the contract documents. Approval by the *Engineer of Record* indicates to the Precaster that he has verified the short and long term adequacy of the structure relative to loads imposed by the precast connections, and that these connections have been checked so that they are adequate and do not interfere with the structural framing.

Drawings sealed by the *Precast Engineer* (engineer who performs the structural design of the precast concrete—may or may not be Engineer of Record) indicates that he takes responsibility for the in-service strength and long-term behavior of the precast, assuming all materials and manufacturing are in accordance with the drawings.

The Architect's and Engineer's approval or corrections and comments on the submitted drawings does not relieve the Contractor of responsibility for any deviation from the requirements of the contract documents unless the Contractor has informed the Architect/Engineer in writing of such deviation at the time of submission and the Architect/Engineer has given written approval to the specific deviation; nor does approval relieve the Contractor from responsibility for errors or omissions in the drawings.

Changes or modifications to the contract documents are not initiated by corrections to the drawings. Written orders (Change Orders) are used to alter the contractual obligations of the Contractor and Precaster.

THE DRAFTING ROOM

Upon receipt of contract documents for a new project, the precast manufacturer's drafting room becomes a nerve center of information and communications. As the individual draftsman translates the design concepts of the project Architect, Engineer of Record, and Precast Engineer into detailed instructions for production and erection, he holds the key to a successful project.

Section IV

Since the draftsman's performance directly will affect the project's success in terms of acceptability and profit, the equipment and organization of the drafting room is seen to be a vital key to success.

The following section is intended to depict the "ideal" drafting room for precast concrete drafting (figure 4-1 illustrates a typical layout). Drafting furniture manufacturers offer *free* advice on laying out drafting rooms.

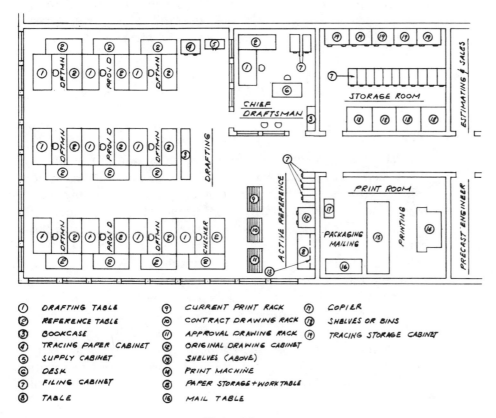

①	DRAFTING TABLE	⑨	CURRENT PRINT RACK	⑰	COPIER
②	REFERENCE TABLE	⑩	CONTRACT DRAWING RACK	⑱	SHELVES OR BINS
③	BOOKCASE	⑪	APPROVAL DRAWING RACK	⑲	TRACING STORAGE CABINET
④	TRACING PAPER CABINET	⑫	ORIGINAL DRAWING CABINET		
⑤	SUPPLY CABINET	⑬	SHELVES (ABOVE)		
⑥	DESK	⑭	PRINT MACHINE		
⑦	FILING CABINET	⑮	PAPER STORAGE + WORK TABLE		
⑧	TABLE	⑯	MAIL TABLE		

Figure 4-1

Investments made to provide the best possible working environment will yield desirable benefits in terms of output, efficiency, and morale; absenteeism due to illness will also be reduced. Janitorial services and regular maintenance should be provided to maintain the working environment.

Eye strain due to inadequate *lighting* can be minimized by providing overhead illumination in the range of 150 to 200 ft. candles. Individual drafting lamps are also recommended to concentrate light on the immediate working area. Windows providing natural light are also highly desirable to eliminate shadows.

Installation of proper *heating* and *air conditioning* equipment is important to avoid temperature extremes within the drafting room. Air conditioning may be required to offset heat generated

Environment

by overhead lights. Dampness and humidity should also be controlled since they have an adverse effect on drafting paper.

Sound is one of the most difficult problems to overcome in the drafting room. To minimize the distracting influence of extraneous noise, installation of sound-absorbing material such as accoustical ceiling tile, drapery, carpeting, and wall coverings is highly recommended. Care in locating sound-emitting machinery should also be exercised.

Electrical outlets should be located at each work area to provide power conveniently for the following items: erasing machine, pencil sharpener, calculator, and drafting lamp.

Personnel Functions and Responsibilities

Each member of the drafting room contributes in his own way to the success of a project. Proper delegation of responsibility based on experience, ability, and position is therefore mandatory. (See figure 4-2.)

The *Chief Draftsman* bears total responsibility for the performance of the drafting room. His duties can be categorized as follows: office administration, bidding and scheduling, monitoring, project administration, and standards' development. His role as *office administrator* encompasses everything from ordering equipment, tools, and supplies to mediating internal personnel problems. When projects are being estimated he is called on to prepare *drafting hour bids* for inclusion in the estimate. To expedite this function he maintains records of past performance of each draftsman and number of man hours required to complete each type of drawing. When a new project is received, he conducts a preliminary review, makes job assignments to the project draftsman, and prepares a *project schedule* for drawing completion (see figure 4-3). This schedule lists projected dates for starting,

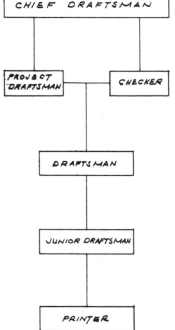

Figure 4-2

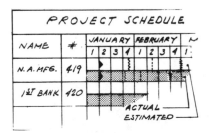

Figure 4-3

submittal of erection drawings, receipt of approval, and completion of production drawings. With many jobs underway at a given time the task of *monitoring* is an important one. This is accomplished ⌐by graphically denoting actual progress versus planned completion on the individual project schedule. This may be accomplished with individual sheets within a notebook or large wall charts.

Project administration is the most time-consuming duty of the Chief Draftsman. He functions as liaison between the drafting room and all other parties concerned. These parties include affiliated departments such as estimating, sales, engineering, accounting, production, shipping, and erection in addition to the Architect, Engineer of Record, Contractor, Owner, subcontractors, suppliers, and building departments. Normally the Chief Draftsman is also charged with the responsibility of developing and maintaining company standards. *Standards' development* is accomplished by constantly reviewing current projects, and selecting for future use details and practices which are desirable from the standpoint of efficiency and economics. All standard company practices and areas of responsibility should be put into writing. This will clearly define the responsibilities of all drafting personnel in complying with instructions. Plant standards should include: (1) procedures to ensure that any changes of details during production or erection are recorded so that the shop drawings always show actual details as finally executed, and (2) quality controls of the detailing process by means of realistic checking procedures. (For a drafting standards manual to be a valuable working tool, it must be used by the draftsmen. If the manual is not kept current, or if its provisions are not adhered to, its production will be merely an unnecessary expense. The staff should be encouraged to follow the standards and to make recommendations for improvement.)

In larger firms the Chief Draftsman may perform little if any actual drafting or layout work. He normally reports directly to, or is an extension of, the precast manufacturer's management.

Obviously the Chief Draftsman cannot personally oversee all aspects of each project. He therefore depends on the *Project Draftsman* to ensure that jobs assigned to him are completed successfully. The Project Draftsman meets with the Chief Draftsman prior to initiating a project and they determine jointly the method of drawing preparation applicable to the specific project, and the potential problems which may arise. The Project Draftsman then determines what portion of the project he will personally execute and assigns the remaining work to a draftsman under his supervision. He also makes daily reports to the Chief Draftsman regarding progress of work currently being done under his direction, coordinates his work with the checker, transmits finished drawings for approval, and releases checked drawings for production.

Responsibility borne by the *checker* includes review of drawings for accuracy and conformance to the estimate, contract documents, precast manufacturer's standards, and Precast Engineer's calculations. He works directly with the Project Draftsman.

when problems are encountered. He marks all of the mistakes and revisions encountered onto a set of prints before returning them to the Project Draftsman. In some cases the checker may also make the physical changes to the drawings.

The responsibility of the *draftsman* is straightforward. He must prepare clear, concise, accurate erection and production drawings under the direction and guidance of the Project Draftsman. He also revises original drawings based on marked-up prints received from the checker or from approval. He reports directly to his Project Draftsman.

Arrangement and Area Requirements

The average drafting room is comprised of from three to twelve employees (Chief Draftsman, Project Draftsmen, draftsmen, and checkers). In addition to providing work areas for personnel and space for printing, active reference and storage areas are required. The following considerations should be kept in mind when planning a drafting room. Each member of the drafting room staff must be provided ample equipment and space to properly function in his capacity. Internal mobility should be afforded all drafting-room personnel. Distractions such as visitors, and sound-emitting machinery should be minimized. Access to the production area should be afforded drafting room personnel. Some privacy should be afforded to individuals within the drafting room. It is desirable to plan for future space requirements in order to maintain a centralized drafting room.

The nature of his assigned duties makes it imperative that the Chief Draftsman be situated in a semi-private or private office. (See figure 4-4.) Many of his conversations, whether in person or

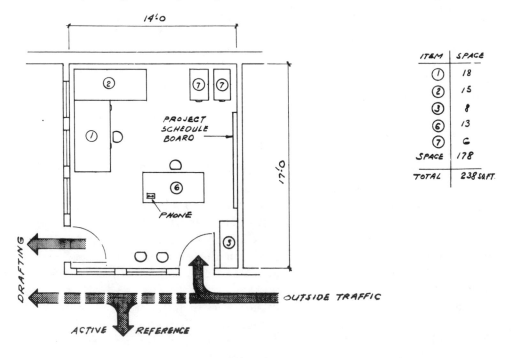

Figure 4-4

by phone, must remain confidential. He should also be able to leave important correspondence and documents on his desk without having them casually observed by others, since they may also be of a confidential nature. A full or partial glass enclosure will permit him to observe the staff and their activities while affording adequate privacy. His office should be situated near the entrance to the department since most traffic entering the department will be to contact him. This location will also enable him to control the flow of traffic at his discretion. His office should contain a desk, drafting table, reference table, book shelf, file cabinets, and chairs. File cabinets are needed for standards, catalogs, and personnel records. Unglazed wall space should also be provided for job status schedules.

To carry out their duties, the Project Draftsman, checker and draftsman require approximately the same arrangement of equipment. (See figure 4-5.) Personnel should be positioned so that they

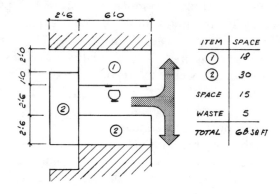

ITEM	SPACE
①	18
②	30
SPACE	15
WASTE	5
TOTAL	68 SQ FT

Figure 4-5

do not face the areas of greatest distraction. They should, where possible, face away from the active job reference area, the main traffic flow, and the Chief Draftsman's office. A drafting table with two reference tables and a drafting stool comprise the main equipment required. Semi-private enclosures may be employed using movable partitions if desired. However, the use of such partitions restrict the Chief Draftsman in his supervisory function. Bookcases, reference shelves, and supply cabinets should be centrally located and readily accessible to all personnel.

An *active reference area* will prove beneficial to even the smallest drafting room. Its purpose is to centrally locate all documentation related to projects in progress. Such material is maintained within this area until project completion at which time it is transferred to the storage area. The primary benefit realized from the establishment of an active reference area is immediate access to project-related information for use during telephone conversations. Figure 4-6 illustrates one arrangement of the equipment required.

The reference table should be 3'0 X 6'0 minimum. This will provide a work surface large enough to handle an opened set of contract drawings. The wall- or table-mounted telephone must be

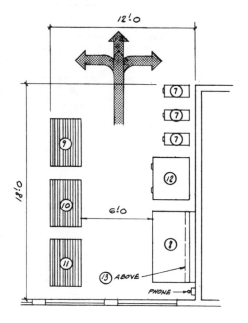

Figure 4-6

equipped with an extension cord of ample length to provide its user complete mobility within the area. Vertical hanging racks are recommended for keeping contract drawings, approval drawings, and current erection/production prints at hand for ready reference. They should have the capacity to receive 30 X 42 inch drawings. A bookshelf or case should be provided for contract specifications and addenda.

Standard size file cabinets are generally employed to retain standard size (8½ X 11 inch) material such as letters, transmittal forms, phone call summaries, memos, sketches, etc. In some cases, legal size (8½ X 14 inch) file cabinets may be required. If the legal size correspondence is not voluminous it is recommended that it be folded to 11 inch length and placed in the standard size cabinet. Drawing cabinets are required for original tracings or they can be kept at the draftsman's desk (tracing drawer in drafting table).

Once *stored*, documentation for a completed project may never be used again. It should, however, be organized in a manner which will make retrieval easy if necessary. When retrieval of stored documents is required, it is normally only for a portion of the material such as a specific letter, one drawing, or a detail from the contract drawings. It is, therefore, practical to store these items separately. Figure 4-1 schematically illustrates one arrangement of inactive storage of documents.

Bins or shelves are recommended to hold rolls of contract and approval drawings. Drawing cabinets ensure that original erection/production drawings are kept in good condition. Specifications are normally stored on bookshelves. Correspondence and calculations are stored in file cabinets. Once this storage system is established, index sheets should be employed to key the location of various items (see figure 4-7).

It is advisable to centralize all *printing* and reproduction operations within the confines of a separate enclosure. (See figure 4-1 for schematic.) The commonly used reproduction machines are the ammonia and dry printers. The ammonia printer requires external ventilation. This arrangement will isolate the noise and fumes emitted during the printing process. In addition to a print machine capable of handling 30-inch wide drawings, a paper storage cabinet or shelf and a work table are required to expedite the printing function. Paper storage drawers or shelves should be labeled to indicate paper size, type, and machine speed. The paper storage unit may also double as a work surface for use during the reproduction process. The print room may also be utilized as a packaging and mailing room. A storage cabinet with work top could hold such items as a postal scale, wrapping paper, cardboard mailing tubes of assorted sizes, stamps, labels, tape, postage meter and/or stamps. Charts for postal zones, rates, and zip codes could also be affixed to the wall.

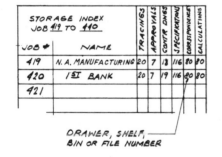

DRAWER, SHELF, BIN OR FILE NUMBER

Figure 4-7

Communications

During the process of drawing preparation, drafting department personnel must engage in effective and precise communications with concerned parties (see figure 4-8). Care in the development of comprehensive procedures and maintenance of complete records is therefore highly recommended. The following is a list of basic rules for various types of communications.

Figure 4-8

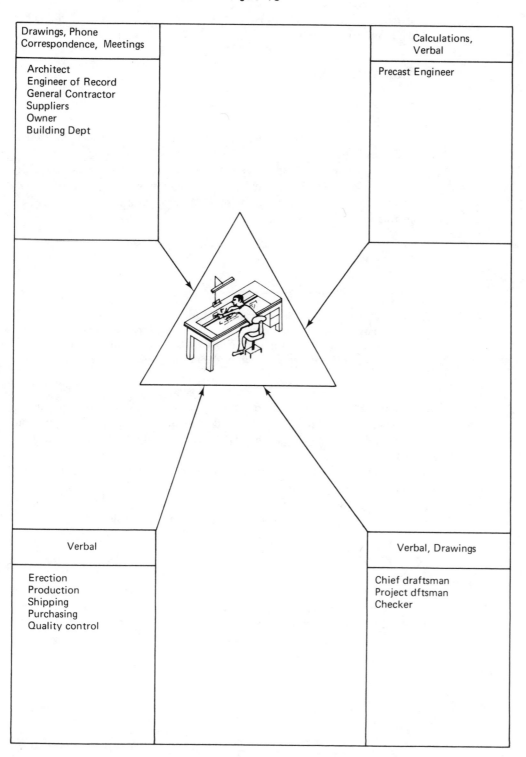

Drawings, Phone Correspondence, Meetings

Architect
Engineer of Record
General Contractor
Suppliers
Owner
Building Dept

Calculations, Verbal

Precast Engineer

Verbal

Erection
Production
Shipping
Purchasing
Quality control

Verbal, Drawings

Chief draftsman
Project dftsman
Checker

1. LETTERS

a. All incoming letters should be date stamped on the day they are received.
b. All correspondence should be reviewed by the Chief Draftsman prior to distribution to other concerned parties.
c. All outgoing letters should be approved and initialed by the Chief Draftsman.
d. Each outgoing letter should contain the date, job name, and number.
e. Always retain file copies of all outgoing correspondence.
f. If for any reason a letter must be removed from the file, a copy should be made and put in its place.

2. TRANSMITTALS

a. Develop and use preprinted forms. (See figure 4-9.)
b. Make duplicate copies for files.
c. All transmittals should be checked and approved by the Chief Draftsman and signed by the Project Draftsman.
d. Transmittals must include: job name and number; sheet number being sent; drawing quantity; drawing type enclosed (sepia, print, original); revision number and date; purpose of transmittal and space for special instructions.

3. PHONE CALLS

> **Phone calls involving seemingly unimportant conversations may eventually become expensive misinterpretations for either or both parties involved. Use of a phone call summary will prevent such problems (see figure 4-10).**

a. Phone calls involving interpretations, clarifications, changes, additions, deletions, etc., to the contract documents, or approved production drawings, *must* be logged by date, time, job number, party called or calling, and subject discussed.
b. Memos *must be written* following conversations and decisions, and *must be checked* and initialed by the Chief Draftsman.
c. A copy of the memo *must be mailed* to the second party as well as to others affected, and the original retained for internal filing.

4. MEETINGS

a. All meetings relative to jobs *must be logged* and memos dispatched using the same procedure as for phone calls.
b. Regular meetings should be held for staff and supervisory personnel to discuss problems, recommend solutions, and participate in planning.

5. VERBAL

a. When communicating job details verbally be as clear and precise as possible.
b. All verbal agreements *must be confirmed in writing.*

To___*J. J. JONES INC.*___
___*112 ELM ST.*___
___*HILLVILLE , ILL.*___

Letter of Transmittal
Architectural Precast Corporation
20 Concrete Avenue
Everytown, Illinois 60202

Date _____ Job No. ___*419*___

Attention___*ART TISHT*___

Gentlemen:

Re:___*N.A. MANUFACTURING*___

We are sending you ___ X ___ Attached _____ Under separate cover via ___*FIRST CLASS MAIL*___

Erection Drawings _____ Production Drawings _____ Prints __X__ Sepias _____ Originals ___

___ Specifications _____ Calculations _____ Other: _____

Copies	Date	No	Description
/	—	E1	ELEVATIONS
/	—	E9	SECTIONS

These are transmitted as checked below:

_X_ For approval   ___ Approved as submitted   Resubmit _____ copies

___ For your use   ___ Approved as noted   for approval
                                                              Submit _____ copies
_X_ As requested   ___ Returned for corrections   for distribution

___ For review & comment   Return _____
                                                              corrected prints

___ For Bids Due _____ 19 _____ Prints Re'd. after Loan to us.

Remarks ___*REMAINDER OF DRAWINGS TO BE SUBMITTED*___
___*NEXT WEEK*___
_____
_____
_____
_____
_____
_____
_____

Copy to   *LEO SPANMORE*_____

Signed   *Al Jouchim*_____

If enclosures are not as noted, kindly notify us at once.

Figure 4-9

The primary function of the drafting department is to furnish written and illustrated instructions to the production shop and to the Contractor for manufacturing and erecting respectively. These instructions are produced in a variety of types and sizes along with other related documents.

Filing and Storage

Architectural Precast Corporation

Date: _____

Time: _3:30 PM_

Phone Call Summary

Outgoing Call ( )

Incoming Call (X)

APC Job: _N. A. MANUFACTURING_   Job No.: _419_

Party(ies) Name: _JOE BESITE_

Party(ies) Firm: _GOTHIC STRUCTURES_

Party(ies) Address: _802 PINE AVE_

HILLVILLE, ILL

Party(ies) Phone: _(318) 354 - 3186_

Subject of discussion: _ERROR IN PLACING L-C-1_

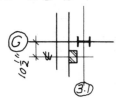

Conclusion: _WE WILL CHANGE
CONNECTION INSERT
LOCATION ON PANEL W 31_

Commitments (if any):   _REVISE SHEET A-5 TO REFLECT "AS
BUILT" CONDITION. RESUBMIT TO ARCHITECT_

(X) Copy to: _ART TISHT_   APC Staff Member: _Al Youchm_
JOE BESITE

Figure 4-10

Maintaining and collecting this material in a systematic manner for each job during its active period, requires an integrated filing system which permits quick and easy retrieval of any part of any job. Keeping different materials in separate files aids in the recovery of any segment of a job.

Development of proper systems of filing and storage can be a tremendous time saver for drafting-room personnel.

The term *filing* as used herein refers to active projects. Figure 4-11 depicts a common procedure for filing active documentation. It is recommended that all filing be concentrated in the active job reference area.

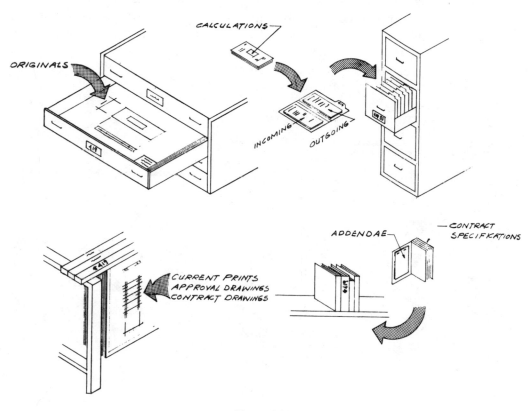

Figure 4-11

1. Erection/production *original* drawings should be filed in horizontal drawers, capable of holding 24 X 36 inch drawings. Drawer fronts should be labeled to indicate job number stored within.

2. *Contract drawings, approval drawings, and current prints* of erection/production drawings should be filed on a vertical hanging rack. Racks should be able to hold 30 X 42 inch drawings. Each "stick" should be labeled on top with the job name and number and hung in numerical order. This method of filing maintains drawings in a flat position which provides better reference than rolls of drawings.

3. *Contract specifications* should be filed on a bookshelf with ends labeled to reflect job number. Addenda should be stapled to the inside front cover.

4. *Correspondence* should be placed in file folders with incoming and outgoing correspondence on opposing covers. This file should also hold calculations prepared by the Precast Engineer. Each folder, labeled with job number, is then filed numerically in standard file cabinets.

The term *storage* as used herein refers to inactive or completed projects. Figure 4-12 depicts procedures for storing inactive documentation upon transfer from the active reference area. The storage of documents no longer required for active reference must be thoroughly organized to provide ready access to each portion of a completed project.

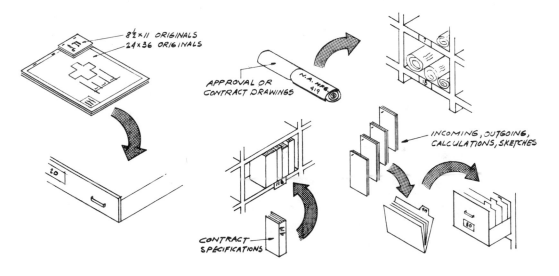

Figure 4-12

1. *Original* erection/production tracings should be placed in numerical order and clamped or stapled together into sets (see figure 2-1). Cabinets which allow flat storage are recommended to maintain the tracings in a printable condition.

2. *Approval drawings* should be rolled tightly, and clearly labeled with job name and number. They should then be stored in shelves or bins with the identification at the exposed end.

3. *Contract drawings* should be stored separately in the same manner as approval prints.

4. Current *prints* of erection/production drawings should be discarded at this point since the originals are already on file.

5. *Contract specifications* are most conveniently stored on shelves. They should be labeled on the visible edge showing job name and number and placed in numerical order.

6. For permanent storage, *correspondence* should be stapled into individual sets of related items. One folder should hold all correspondence and calculations. All correspondence should be placed in chronological order according to date written. Folder tab should be labeled with job name and number and stored in numerical order.

An effective method of reducing the volume of stored material is the use of microfilm. Microfilming of contract documents and original erection/production drawings can save as much as 80 percent in storage space. Disposal of bulky rolls of drawings becomes possible once the information is recorded on microfilm.

Information on the entire range of basic and sophisticated drafting equipment can be found in commercial drafting supply catalogs.

Equipment, Tools, and Supply Glossary

Drafting Table. Minimum specifications are as follows: (See figure 4-13.)

width: 36"
length: 72"
height: 36"
construction: wood or steel
work surface: vinyl covered

drawer space: one tracing storage drawer, one tool storage drawer, and one file drawer
attachments: drafting lamp and parallel straight edge.

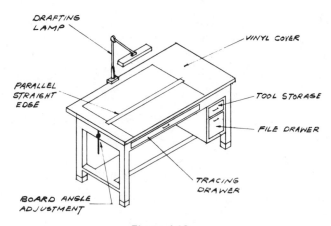

Figure 4-13

Drafting Chair. Minimum specifications are as follows:

> seat height: adjustable from 17 to 34 inches
> back rest height: adjustable from
> 10 to 14 inches above seat
> frame construction: steel
> seat cushion and back rest construction:
> vinyl upholstery
> additional desirable features:
> swivel top and caster mounted base

Reference table. Minimum requirements are as follows:

> width: 30″   construction: steel or wood
> length: 72″   features: storage shelf
> height: 30″   below additional drawer space

The following tools should be supplied to individual draftsman:

Scales Civil engineering scale divided into decimal parts of 10, 20, 30, 40, 50, 60 and 80 division to the inch.
Architectural scale divided into proportional feet and inch divisions of ⅛, ¼, ⅜, ½, ¾, 1, 1½, and 3 inches to the foot. (Principle scale used for precast concrete drawings.)

Triangles An 8 inch high 45° and a 10 inch high 30/60° will prove most useful. Colored triangles tend to refract light and eliminate troublesome shadows along their edges. An adjustable triangle which combines the function of a triangle and a protractor is also recommended.

Templates Plastic templates which simplify construction of geometric figures and other symbols are a tremendous time saver.

Slide Rule and/or Calculator These tools are an absolute must to facilitate rapid calculation of volumes, weights, areas, and centers of gravity which are essential parts of the draftsman's work.

Compass A hand compass is required to construct curves, arcs, and circles which, due to their size, cannot be drawn by template.

Supplies which should be furnished to individual draftsman include:

Mechanical pencil or lead holders   Drafting leads in hard, medium, and soft grades

Lead pointer

Electric eraser   Drafting brush

Erasing shield   Sepia Eradicator

Supplies which should be available as part of the general office inventory include:

Adjustable curve	Paper punch
Drafting paper	Paper cutter
Beam compass	Stapler
Paper shears	Staple remover

Lettering instruments

APPENDICES

Prestressed Concrete Institute, Chicago, Illinois

1.* PCI Architectural Precast Concrete, 1973

2.* PCI Design Handbook—Precast and Prestressed Concrete, 1971

3. PCI Manual on Design of Connections for Precast Prestressed Concrete, 1973

4. Manual for Quality Control for Plants and Production of Architectural Precast Concrete Products, 1975

5. Veltman, C. J. and Johnson, R. W., *Effective Shop Drawing Communications for Precast Concrete*, Journal of the Prestressed Concrete Institute, Vol. 14, No. 1, February 1969, pp. 12–31

6. Raths, C. H., *Production and Design of Architectural Precast Concrete*, Journal of the Prestressed Concrete Institute, Vol. 12, No. 3, June 1967, pp. 2–27

7. Architectural Precast Concrete Joint Details, PCI Committee on Architectural Precast Concrete Joint Details, Journal of the Prestressed Concrete Institute, Vol. 18, No. 2, March-April 1973, pp. 10–37

American Concrete Institute, Detroit, Michigan

8. ACI Manual of Concrete Inspection, SP-2, 1967

9. Manual of Standard Practice for Detailing Reinforced Concrete Structures, 1974

American Welding Society, Miami, Florida

10. AWS D1.0—Code for Welding in Building Construction

11. AWS D 12.1—Recommended Practices for Welding Reinforcing Steel, Metal Inserts and Connections in Reinforced Construction

American Society for Testing and Materials, Philadelphia, Pa.

12. Compilation of ASTM Standards in Building Codes, 1973

Portland Cement Association, Skokie, Illinois

13. Design and Control of Concrete Mixtures, 1968

---

*Strongly recommended to provide supplementary information to the contents of this manual.

Concrete Reinforcing Steel Institute, Chicago, Illinois

14.* Manual of Standard Practice, 1973

15. CRSI Design Handbook, 1972

Wire Reinforcement Institute, Inc., McLean, Virginia

16. Manual of Standard Practice Welded Wire Fabric, October, 1972

American Institute of Steel Construction, New York, N.Y.

17.* Manual of Steel Construction, 1970

18. Structural Steel Detailing, 1971

Non-Association Publications

19. *Structural Engineering Aspects of Headed Concrete Anchors and Deformed Bar Anchors in the Concrete Construction Industry.* KSM Welding Systems, Omark Industries, SA 1-KSM 2017-5-127 4, 1974

20. Ramsey, C. T. and Sleeper, H. L., *Architectural Graphic Standards,* Wiley & Sons, 1970

22. Gage, M., *Guide to Exposed Concrete Finishes*, Cement & Concrete Association, London, 1970

22. Callender, J. H., *Time-Saver Standards: A Handbook of Architectural Design*, McGraw-Hill Book Co., 1966

23. Fairweather, L. and Sliva, J.A., *VNR Metric Handbook* Van Nostrand Reinhold Co., 1969

24. Architectural Catalog File, Sweet's Construction Division, McGraw-Hill Information System Company, Yearly

25. Smoley's Four Combined Tables, C. K. Smoley and Sons, Inc., Chautauqua, N.Y.

26. Building Codes—Local

27. Grachino, J. W., Weeks, W., and Brune, E., *Welding Skills and Practices*, American Technical Society, 1971

28.* Lincoln Electric Stick Electrode Welding Guide, Lincoln Electric Company, 1972

*Strongly recommended to provide supplementary information to the contents of this manual.

SOURCES OF ADDITIONAL INFORMATION

SUBJECT	REFERENCE
Architectural Standards	20, 22
Concrete (Fundamentals)	8, 13
Concrete (Inspection and Testing)	4, 8, 13
Concrete (Precast)	1, 2, 3, 4, 6, 7
Concrete (Properties)	8, 13
Concrete (Reinforcing)	14, 15
Drafting (Precast)	5
Drafting (Reinforced Concrete)	9
Drafting (Structural Steel)	18
Finishes	1, 4, 21
Joints	1, 7
Material (Information)	24
Material (Requirements)	26
Material (Specifications)	12
Mathematics	25
Metric System	23
Rebar	14
Structural Steel	17
Studs (Headed and Deformed)	19
Tolerances (Precast)	1, 4
Welded Wire Fabric	14, 16
Welding	10, 11, 27, 28

It is in the interest of all parties—Prime Consultants, Contractors, and Precasters—to have a clear understanding of the duties and responsibilities of each party (in a general sense) to avoid confusion of roles and to promote a more uniform working relationship.

Responsibility
of Prime Consultant

The responsibilities of the Prime Consultant (Architect and/or Engineer of Record) are subject to his contractural relationship to the Owner and also to the rights, duties, obligations, and responsibilities assigned to him in the contract documents. Exceptions to items of responsibility are clarified in a written instrument to which the Owner and Contractor are parties, and in the Owner/Prime Consultant agreement.

Responsibility for both structural and aesthetic design of architectural precast concrete should rest with the Prime Consultant who should:

1. Provide clear and concise drawings and specifications and, where necessary, interpretation of the contract documents.

2. Establish the standards of acceptability. The Prime Consultant's right to reject work is provided in order to enable him to fulfill his duty to the Owner in endeavoring to guard the Owner against defects in the work which will render the building, or any portion of it, incapable of use in the manner and for the purpose for which it was intended, or from having the intended appearance.

3. Determine the part, if any, played by the precast concrete in the support of the structure as a whole.

4. Provide as part of his design for the effect of difference in material properties, stiffness, temperatures, and other elements which might influence the interaction of precast concrete units with the structure.

5. Determine load reactions necessary for the accurate design of both reinforcement and connections.

6. Evaluate thermal movements as they might affect requirements for joints, connections, reinforcement, and compatibility with adjacent materials.

7. Analyze the water-tightness of exterior concrete wall panel systems; evaluating joint treatment, including the performance of adjacent materials for compatibility in joint treatment and

the proper sealing of windows and other openings, except where such systems are manufactured and marketed as a proprietary item.

8. Design the supporting structure so that it will carry the weight of the precast concrete as well as any superimposed loads including provisions for deflection and rotation of supporting structure during and after erection of the precast concrete.

9. If the stability of the structure is dependent on the sequence or method of erection, design the supporting structure to withstand the temporary loading conditions that might be encountered as a result of the sequence of erection and/or the sequence of loading of the structure. However, the Contractor is responsible for construction means, methods, techniques, sequences, or procedures.

10. Make selection of surface finishes recognizing certain limitations in materials and production techniques in regard to uniform color, texture, and performance, especially the limitations which are inherent in natural materials.

11. Make selection of interior finishes defining the area of exposure and the interior appearance for occupancy requirements; again recognizing material and production limitations.

12. Design for durable exterior walls with respect to weathering, corrosive environments, heat transfer, vapor diffusion, and moist air or rain penetration.

13. Design sandwich wall panels for their proper structural and insulating performance.

14. Review and approve all manufacturers' drawings as required in the specifications and in keeping with Section III, The Submittal Process.

15. Specify dimensional and erection tolerances for the precast concrete, and tolerances for supporting structure and contractors' hardware. Any exceptions to PCI standard tolerances (Appendix E) should be clearly identified.

Contract drawings should provide a clear interpretation of the configuration and dimensions of individual units, their relationship to the structure, and to other materials. The drawings should indicate the location of all joints, both real (functional) and false (aesthetic). The drawings must supply the following information:

1. All sections and dimensions necessary to define the size and shape of the unit

2. Joint details between units
3. Finish required on all surfaces and a clear indication of the surfaces that are to remain exposed
4. Details of corners of the structure
5. Details of connections to the supporting structure, optional
6. Reinforcement of units, optional
7. Jointing to other materials
8. Unusual conditions

This information should be sufficiently detailed to enable the Precaster to produce the units and for the erector to install them. In most instances the Precaster will make his own erection drawings to identify the information for precast concrete and translate these details into production and erection requirements.

Whether all reinforcement and connection details are shown by the designer or left entirely to the Precaster is the prerogative of the Prime Consultant. This condition may be determined by local practices and included in the project specifications. If reinforcement and connections are not detailed, the requirements for the design must be clearly identified. The amount of space allowed for connections should also be indicated.

It is generally recommended that the Prime Consultant— Architect or Structural Engineer—design the reinforcement and the connection for each typical unit. This approach provides a design compatible with structural capability, concrete cover, hardware protection, appearance, and clearance for mechanical services. In addition it will establish parameters for modifications which may be suggested by the Precaster to suit his production and erection techniques, and at the same time satisfy project design requirements. Design deviations requested by the Precaster or Erector will be permitted only after the Architect/Engineer's approval of the proposed change.

Responsibility of Constructor

The General Conditions of the contract state the responsibility of the constructor (usually the General Contractor) in coordinating the construction work. The General Contractor is responsible for project schedule, dimensions and quantities, coordination with all other construction trades, and for the adequacy of construction means, methods, techniques, sequences, and procedures of construction, in addition to safety precautions and programs in connection with the project. The precast manufacturer should not proceed with fabrication of any products prior to receiving approval of erection drawings by the General Contractor. The constructor should:

1. Be responsible for coordinating all information necessary to produce the precast erection drawings.
2. Review and approve all precast erection drawings which include the scheme of handling, transporting, and erecting the

precast concrete, as well as the plan of temporary bracing of the structure. (Handling and transporting responsibilities depend on whether precast is sold F.O.B. plant or jobsite.)

3. Be responsible for the coordination of dimensional interfacing of precast with other trades.

4. See that proper tolerances are maintained to guarantee accurate fit and overall conformity with precast erection drawings.

The constructor must be responsible for coordinating complete precast erection drawings so that related items are transmitted by him to the designer in one package. The constructor must immediately notify the Precaster of any deviations found in dimensions due to plan or construction errors.

Responsibility of the Precast Concrete Manufacturer

The manufacturer should review the design of the precast concrete elements for structural soundness and feasibility with respect to finishes, connections, handling stresses, material quality, joint treatment, and tolerances for both manufacturing and installation. The manufacturer should report any discrepancies to the Prime Consultant (preferably prior to bidding). This does not relieve the Prime Consultant of his responsibilities as outlined before.

The manufacturer should analyze all precast units for handling or temporary loadings imposed on them prior to and during final incorporation into the finished building or structure. He should design and provide reinforcement or temporary strengthening of the units to ensure that no stresses are introduced which will exceed the design requirements of codes or standards governing the project, or have an adverse effect on their performance or safety after installation.

Provisions for any construction loads which are in excess of stated design requirements, and which may occur after installation of the units, are not the responsibility of the manufacturer, (General Contractor's responsibility).

Any additional design responsibility vested with the precast concrete manufacturer should be clearly defined by the Prime Consultant. Additional design responsibility for the manufacturer may occur when the Prime Consultant uses methods of communication as listed in Options II and III in Table B-1.

Most precast work today is covered in Option I. Option II (b) may occur when no engineer is involved with the Architect for panel design, and where doubtful areas of responsibility therefore may exist. Under this Option, the manufacturer should employ or retain a structural engineer experienced in the design of precast concrete, who will ensure the adequacy of those structural aspects of the erection drawings, manufacture and installation for which the manufacturer is responsible. Option III is not yet a common practice but might be used for design of systems buildings.

Table B-1	DESIGN RESPONSIBILITIES
Contract Information Supplied by Prime Consultant:	Responsibility of the Manufacturer:
OPTION I	
Provide complete drawings and specifications detailing all aesthetic and functional requirements plus dimensions.	The manufacturer shall make erection and production drawings (as required) with details as shown by the prime consultant. He may suggest modifications that in his estimation would improve the economics, structural soundness or performance of the precast installation. The manufacturer shall obtain specific approval for such modifications. Full responsibility for the precast design, including such modifications, shall remain with the prime consultant.
OPTION II	
Detail all aesthetic and functional requirements but specify only the required structural performance of the precast units. Specified performance shall include all limiting combinations of loads together with their points of application. This information shall be supplied in such a way that all details of the unit can be designed without reference to the behavior of other parts of the structure.	The manufacturer has two alternatives (a) Submit erection and shape drawings with all necessary details and design information for the approval and ultimate responsibility of the prime consultant. (b) Submit erection and shape drawings for general approval and assume responsibility for his part of the structural design, i.e., the individual units but not their effect on the building. Firms accepting this practice may either stamp (seal) drawings themselves, or commission engineering firms to perform the design and stamp the drawings. The choice between alternatives (a) and (b) shall be decided between the prime consultant and the manufacturer prior to bidding with either approach clearly stated in the specifications for proper allocation of design responsibility.
OPTION III	
Cover the required structural performance of the precast units as in Option II and cover all or parts of the aesthetic and functional requirements by performance specifications. Define all limiting factors such as minimum and maximum thickness, depths, weights and any other limiting dimensions. Give acceptable limits of any other requirements not detailed.	The manufacturer shall submit drawings with choices assuming responsibilities as in Option II.

The efficiency of any communications system—drawings or specifications—lies in the ability of all concerned to understand the terminology used in the documents.

Since the range of individuals who can benefit from a glossary is so wide, this is an attempt to make definitions as basic as possible. It is not intended to cover every term used in the precasting industry, but those which are most used or misused. Every effort has been made to include all terms common to drafting, engineering, production, and erection of precast concrete. Some definitions may be at variance with the most commonly accepted meaning due to regional variations.

Figure C-1 illustrates some of the terms used in this handbook, others are defined below.

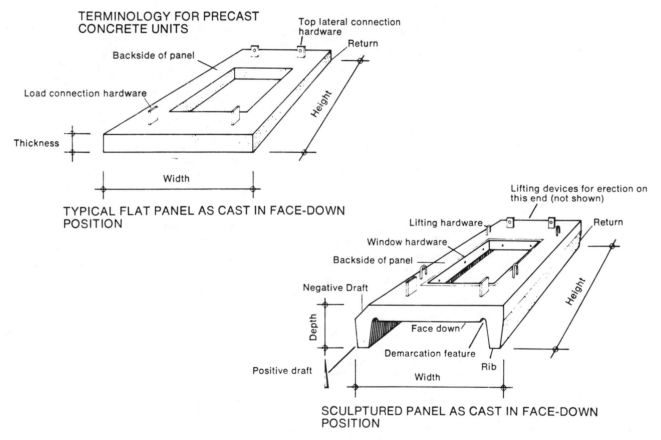

Figure C-1

A

abrasive nosing. A non-skid metal unit which is cast into the nose of a precast stair panel.

absorption. The process by which water is absorbed; the amount of water absorbed under specific conditions, usually expressed as percentage of dry weight of the material.

ABRASIVE NOSING

absorption test for concrete. A test for early indication of predictable weather staining (rather than durability).

addendum. A supplement to specifications or contract drawings issued prior to the execution of the construction contract.

admixture. A material other than water, aggregates, and cement used as an ingredient of concrete or grout to impart special characteristics.

admixture, air entraining. A material added to the concrete for the purpose of entraining minute bubbles of air in the concrete during mixing and thus improving the durability of concrete exposed to cyclical freezing and thawing in the presence of moisture.

"A" FRAME

A frame. An A-shaped frame used to support panels on flat bed trucks during shipping.

aggregate. Naturally occurring, processed or manufactured, inorganic particles which are mixed with portland cement and water to produce concrete; normally comprises 60 to 80 percent of the total volume of concrete.

aggregate, coarse. Aggregates with particle sizes greater than about 1/4 inch.

aggregate, fine. Natural or manufactured sand with particle sizes smaller than about 1/4 inch.

aggregate, structural lightweight. Aggregate having a dry, loose weight of 70 PCF (pounds per cubic foot) or less.

aggregate transfer. A method of obtaining an exposed aggregate surface; aggregates are held in an adhesive on form liners, the liners are installed in the forms, concrete is placed and cured, and forms and liners are removed; aggregates become embedded in and bonded to the concrete to such an extent that they are transferred from the liners to the concrete.

aircraft cable. Multi-strand steel cable, in loop form, cast in precast panels for handling purposes; cable is more flexible than prestressing strand.

air entrainment. An increase in the amount of air in a concrete mix through the use of an air-entraining admixture (see admixture, air entraining); air entrained concrete displays increased workability and cohesiveness.

air pocket. Pits (entrapped air or water bubbles) in the form faces of a panel caused by improper consolidation or inadequate draft.

alignment face. Face of a wall panel which is to be set in alignment with the face of adjacent panels.

ANCHORS

alternate. A method or material to be used in place of that originally shown.

ambient temperature. Temperature of the surrounding air and of the forms into which concrete is to be cast.

anchor. (1) Headed studs, deformed studs, straps, rebar, etc. welded to steel angles or plates and embedded in concrete for use as part of a connection. (2) any item cast into or pre-affixed to the structure for the purpose of receiving a connection.

angle iron. A steel section consisting of two legs at an angle (which is almost invariably a right angle).

ANGLE

approval. Acceptance of the Precaster's drawings by the Architect, Engineer of Record, and General Contractor indicating that all building conditions and dimensions shown are correct and final.

architect. A person or firm that determines the general design and appearance of a building, draws the plans, writes the specifications, and supervises the construction.

architectural precast concrete. Any precast concrete unit of special or occasionally standard shape that through application or finish, shape, color, or texture contributes to the architectural form and finished effect of the structure; units may be structural or decorative, and may be conventionally reinforced or prestressed.

arc welding. A process by which two pieces of steel to be joined are heated by an arc formed between an electrode and the steel; as the electrode melts, it supplies weld material which fuses the pieces of steel together.

area of steel. Cross-sectional area of reinforcing bars required for a given concrete section.

arris. Shape edge or ridge formed by two surfaces meeting at an angle.

assembly. A set of parts arranged into one unit.

axial load. A load applied on the axial center of an element.

BACKSPAN

B

back span. The distance between the supports of a cantilevered member.

backup material. Material used to limit the depth of the sealant in panel joints.

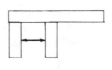

BACKUP MATERIAL

backup mix. The concrete cast into the mold as a filler behind a thin layer of the more expensive face mix.

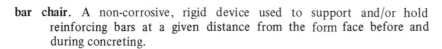

BACKUP MIX

bagtie. Thin gage wire ties (generally No. 16, 15, or 14 gage) used to fasten reinforcing bars together at intersections.

balance point. (See center of gravity.)

bar chair. A non-corrosive, rigid device used to support and/or hold reinforcing bars at a given distance from the form face before and during concreting.

BAR CHAIR

bar joist. A lightweight steel truss for supporting floors and roofs.

bar number. A number (approximately the reinforcing bar diameter in eighths of inches) used to designate reinforcing bar size; bar numbers are rolled onto the bar for easy identification.

BAR JOIST

bar stool. (See bar chair.)

base line. The bottom reference point of a building which serves as a basis for measurement and/or dimensions.

base plate. A steel plate anchored to the bottom of a precast unit for the purpose of fastening it to the foundation.

BASE PLATE

batch. The amount of concrete produced in one mixing operation.

batter. (See draft.)

battery mold. A series of reuseable casting slots used to manufacture panels in a vertical position, thus reducing reinforcing requirements and imparting a smooth form finish to both faces.

bay. Square or rectangular areas, usually in a uniform grouping which are contained between adjacent columns, beams, and/or walls.

beam. A horizontal structural member carrying loads from a floor or roof system (which induce bending) and usually spanning between columns, girders, piers, or walls.

beam pocket. Opening in a vertical member in which a beam is to rest.

bear. To transfer vertical load to another member.

bearing area. The surface in square inches, which comes in contact with a vertical load transferring member.

BEARING PAD

bearing pad. A pad, usually neoprene, which is placed between a member and its support.

BEARING PLATE

bearing plate. A steel- or teflon-coated plate placed between a member and its support.

belt course. A flat, horizontal panel which bands the perimeter of a building marking a division in the wall plane.

bench mark. A datum point, the elevation of which is known, from which differences in elevation are determined.

bent. For analysis purposes, the structural cross-section through a bay of a building.

bill of materials. Material list for individual project (see page 71).

bituminous paint. Paint made from bituminous coal by-products used to prevent corrosion of steel; applied after all welding is completed.

BLEED HOLE

bleed hole. A hole in a plate or angle which is provided solely to release entrapped air or water during concrete placing operation.

bleeding. A form of segregation in which some of the water in a mix rises to the surface of freshly placed concrete; also known as water gain.

blocking. The shims required to level and/or plumb a unit in its proper position.

BLOCKING POINTS

blocking points. Two predetermined locations at which a panel is to be supported during storage and/or shipping to minimize bending moments.

blockout. To form a hole, or reduce the height or width of a panel by affixing material to the form (space within a form in which concrete is not to be placed). (See page 178.)

blowhole. Approximately ½-inch diameter cavities in the form face of a panel caused by an adhering air or water bubble not displaced during consolidation (see air pocket).

blueprint. A nontransparent white on blue copy of a drawing.

bond. Adhesion of concrete to reinforcement or to other surfaces against which it is placed.

bond beam. A horizontal reinforced concrete masonry member

BOND BEAM

bond breaker. A substance placed on a material to prevent it from bonding to the concrete, or between a face material such as natural stone and the concrete backup.

bonding agent. A substance used to increase the bond between an existing piece of concrete and a subsequent application of concrete such as a patch.

BOWING

bowing. The deflection of a vertical panel in a single plane.

break line. Lines used to increase or decrease a large item or section.

bridging. Braces between and perpendicular to steel floor and roof members which provide stress distribution.

BRIDGING

bughole. (See blowhole.)

building code. Laws or regulations set up by building departments of cities, states, and Federal Government for uniformity in construction, design, and building practices.

bulkhead. A vertical partition in the form blocking fresh concrete from a section of the form; divides a continuous casting bed into given unit lengths.

bundling. Placing several parallel elements of reinforcement in contact with each other.

burr. (1) To disfigure the threads of a bolt by stroking with a screwdriver or other tool; (2) Unwanted uneven projections on the edge of a precast panel.

bush-hammer. To break the smooth surface finish of a panel with a pneumatic tool having a serrated face.

by others. Service or material supplied and/or installed by someone other than the precast manufacturer.

C

cadmium plated. Electroplated with noncorrosive cadmium.

cage. Reinforcing bar assembly comprised of rebar and/or welded wire fabric.

calculations. Neat, concise, numerical justification for reinforcement, panel sizing, connections, etc. prepared by a structural engineer.

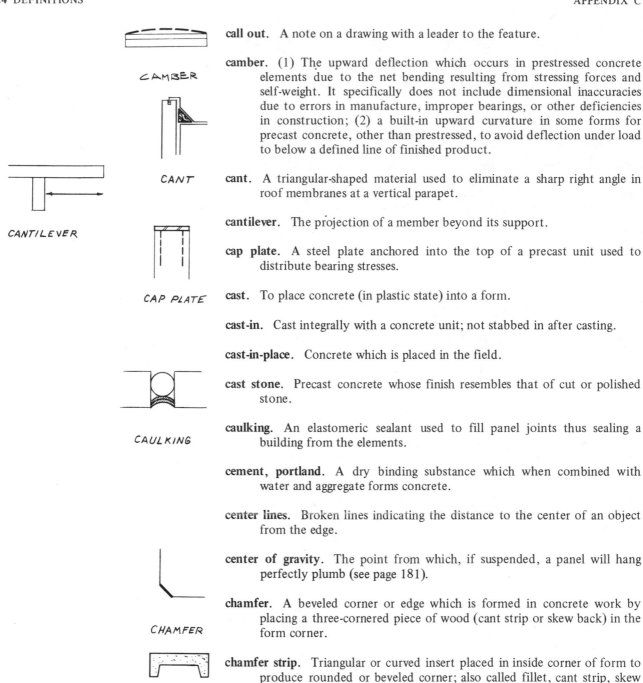

CAMBER

CANT

CANTILEVER

CAP PLATE

CAULKING

CHAMFER

CHANNEL

call out. A note on a drawing with a leader to the feature.

camber. (1) The upward deflection which occurs in prestressed concrete elements due to the net bending resulting from stressing forces and self-weight. It specifically does not include dimensional inaccuracies due to errors in manufacture, improper bearings, or other deficiencies in construction; (2) a built-in upward curvature in some forms for precast concrete, other than prestressed, to avoid deflection under load to below a defined line of finished product.

cant. A triangular-shaped material used to eliminate a sharp right angle in roof membranes at a vertical parapet.

cantilever. The projection of a member beyond its support.

cap plate. A steel plate anchored into the top of a precast unit used to distribute bearing stresses.

cast. To place concrete (in plastic state) into a form.

cast-in. Cast integrally with a concrete unit; not stabbed in after casting.

cast-in-place. Concrete which is placed in the field.

cast stone. Precast concrete whose finish resembles that of cut or polished stone.

caulking. An elastomeric sealant used to fill panel joints thus sealing a building from the elements.

cement, portland. A dry binding substance which when combined with water and aggregate forms concrete.

center lines. Broken lines indicating the distance to the center of an object from the edge.

center of gravity. The point from which, if suspended, a panel will hang perfectly plumb (see page 181).

chamfer. A beveled corner or edge which is formed in concrete work by placing a three-cornered piece of wood (cant strip or skew back) in the form corner.

chamfer strip. Triangular or curved insert placed in inside corner of form to produce rounded or beveled corner; also called fillet, cant strip, skew back.

channel. A precast unit having the shape of an open rectangle.

chase. (1) A vertical space within a building for ducts, pipes, or wires; (2) A long groove or recess formed or cut in panel.

chert. Unsound aggregates which are subject to considerable change in volume and can result in surface pop-outs.

chocker. A chain or cable sling hitch used for handling or lashing down precast units; the greater the load the tighter the grip.

chuck. A device which locks strands into their elongated position during the prestressing operation.

cinch anchor. A type of expansion bolt.

clearance. The distance between two surfaces.

clear span. The distance between the inside edges of the bearing surfaces of two supporting members.

clevis. A U-shaped piece of metal with holes in each end through which a pin is placed.

clip angle. A noncontinuous steel angle used for fastening precast units.

coil thread. A helical-shaped thread which fits the contour and diameter of the wire from which a coil insert is formed. This thread is fast, nonclogging, self-cleaning, and damage-resistant.

cold joint. A joint necessitated by several casting stages but designed and executed to allow the separate components to appear and perform as one homogeneous unit; term only applies when the first casting is allowed to harden prior to placing the second.

column. An element used primarily to support axial compressive loads and with a height at least three times its smallest lateral dimension.

column cover. A precast panel which covers one or more sides of a column which would otherwise be exposed.

column line. A plan reference line which is usually the centerline or exterior face of a column as determined by the architect.

come along. A chain- or cable-type erection device used to bring a panel into the building once the base is set and the crane has been released.

composite construction. A type of construction wherein the floor slab is fastened to the beams in such a manner that they act together as a more efficient member in carrying live loads.

compression. A force which works to compress, compact, and shorten.

compressive strength. The measured resistance of a concrete specimen to axial loading expressed as pounds per square inch (PSI) of cross-sectional area. The maximum compressive stress which concrete, or grout is capable of sustaining.

concrete. A mixture of portland cement, fine aggregate, coarse aggregate, and water.

concrete, structural lightweight. Concrete that has a 28-day compressive strength in excess of 2,500 PSI and an air-dry unit weight of less than 115 PCF.; a lightweight concrete without natural sand is termed *all-lightweight concrete* and lightweight concrete in which all fine aggregate consists of normal weight sand is termed *sand-lightweight concrete.*

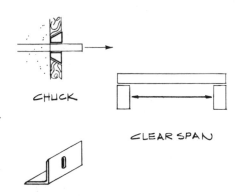

CHUCK

CLEAR SPAN

CLIP ANGLE

COLUMN COVER

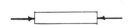

COMPRESSION

concrete cover. The clear distance from the face of the concrete to the reinforcing steel.

connections. Devices for the attachment of precast units to each other or to the building structure.

consistency. The degree of plasticity of fresh concrete; the normal measure of consistency is slump.

consolidation. The use of hand tools, vibrators, or finishing machines during the casting process to eliminate voids, other than entrained air, and to provide a dense concrete, good bond with reinforcement, and a smooth surface.

continuous graded mix. A concrete mix that contains all the sizes of aggregates (below a given maximum) in amounts which ensure an optimum density of the mix.

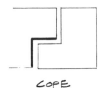

COPE

contract documents. Architectural drawings, structural drawings, specifications, addenda, etc. from which projects are bid and precast drawings prepared.

cope. To cut away a portion of one member to provide clearance for another member.

coping. A panel which forms the top of a wall and seals it from the elements.

COPING

corbel. Steel or reinforced concrete bracket which protrudes from a column or wall panel to provide support for another member or to take support from an adjacent part of structure.

CORBEL

core drill. To cut holes in concrete with a cylindrical bit.

cornice. Panel which fits under a ceiling or projecting roof.

coursing. Module which coincides with the even vertical spacing of brick or block.

CORNICE

covermeter. (See pachometer.)

crazing. A network of fine cracks in random directions breaking the exposed face of a panel into areas of from ¼ inch to 3 inches across. Some probable causes are rich cement mix, too early stripping, and inadequate curing.

creep. The long-term non-elastic shortening of concrete in compression under sustained load.

cross hatch. Lines drawn closely together, generally at an angle of 45 degrees, to denote a sectional cut.

crust. A thin layer of dried cement paste which forms on the top surface of a panel as a result of additional water being added during placing.

cull. A precast unit which is discarded for reasons of imperfection.

curing. The maintenance of humidity and temperature of freshly placed concrete during some definite period following placing, casting, or finishing to assure satisfactory hydration of the cementitious materials and proper hardening of the concrete; where the curing temperature remains in the normal environmental range (generally between 55F and 90F) use the term **normal curing**; where the curing temperature is increased to a higher range (generally between 90F and 150F) use the term **accelerated curing**.

curtain wall. Precast wall panels which when in place may form window frames and interior/exterior wall construction, but support no loads from building.

cut sheet. Small drawings, usually 8½ inches x 11 inches, showing details of individual assemblies or panel conditions.

D

deflection. The distance a structural member moves from its normal position when subjected to a load.

DEFLECTION

deformation. A change of dimension or shape in a body resulting from external loading.

deformed bar. Reinforcing bar manufactured with deformations (bumps, lugs, or ridges) to provide a locking anchorage (bond) with the surrounding concrete.

deformed stud. A steel rod having an irregular surface texture and used for anchoring an angle or plate in concrete; stud is fastened to angle or plate with a special gun (stud welder) by passing a current through the stud.

DOVETAIL ANCHOR

detail. An enlarged drawing of an area which would be difficult to understand at its previous scale.

dimension line. Fine lines placed outside the view or object to indicate measured distance.

DOVETAIL CONNECTION

dovetail anchor. A tenon consisting of a flat piece of light-gage metal.

dovetail connection. A connection system which employs a mortise and tenon interlock joint.

DOVETAIL SLOT

dovetail slot. A mortise consisting of a pre-formed slot.

dowel. A metal pin used to fasten panel bases to cast-in-place concrete or other precast, by fitting into corresponding holes in the respective units; dowel type connections are a carry-over from stone construction and should not be used indiscriminately.

DOWEL

draft. The slope of concrete surface in relation to the direction in which the precast element is withdrawn from the mold; it is provided to facilitate stripping with a minimum of mold breakdown.

DRAFT

draw. (See draft.)

DRIP

drip. A projecting fin or groove at the outer edge of a sill, projecting horizontal wall element or soffit, designed to interrupt the flow of rainwater downward over the wall or inward across the soffit; drips are normally used only on units having a smooth or lightly exposed finish.

dry-pack. Hand grouting with a very dry mix; the grout is tamped into the joint.

ductility. The property of a material to stretch or "give a little" when overloaded rather than rupture.

dunnage. Materials used for keeping concrete elements from touching each other or other materials during storage and transportation.

durability. The ability of concrete to resist weathering action, chemical attack, and abrasion.

durometer. The indication of hardness of a material (i.e., neoprene pads); higher numbers indicate increased hardness.

ECCENTRICITY

E

eccentricity. The distance between the center of a load and the center of its support.

effective length. The length used for design of compression members.

efflorescence. A crystalline deposit of soluble salts, usually white in color, appearing on the surface of concrete; salts are carried in solution to the surface of the concrete where carbonation and evaporation take place; incidence of efflorescence is largely regulated by permeability and texture of the concrete surface; some probable causes of efflorescence are high water cement ratio, poor release agent application and non-uniform curing; dense concrete with low absorption is less susceptible to this condition.

elevation. (1) The dimension (above (+) positive, below (−) negative) from any point to the established datum; (2) Drawing of front, sides, or rear face of a building in a vertical plane, usually made as though the observer were looking straight at it.

END CLOSURE

elongation. In prestressed work, the difference between a strand's initial length and its length after stressing.

end closure. A precast unit which fits between the stems of a prestressed slab forming a diaphragm or wall closure.

end rail. That portion of a form which dictates the top and bottom of a panel.

engineer of record. Engineer who creates original building design and is responsible for the design.

envelope mold. A box mold where all sides remain in place during the entire casting and stripping cycle.

erection. The placing of precast units into their respective positions in the structure.

erection drawings. (See page 20.)

erection mark. An identification mark or number placed on the end of each member to aid in the erection of the structure.

expanded metal. A light gage flat metal with large holes occurring in a regular pattern.

EXPANDED METAL

expansion bolt or anchor. An expandable device made of metal inserted into a drilled hole in hardened concrete that grips concrete by wedging action when the nut or head is rotated.

exposed aggregate concrete. Concrete with the aggregates exposed by surface treatment. Different degrees of exposure are defined as follows:

Light exposure—where only the surface skin of cement and sand is removed, just sufficient to expose the edges of the closest coarse aggregate.

Medium exposure—where a further removal of cement and sand has caused the coarse aggregate to visually appear approximately equal in area to the matrix.

Deep exposure—where cement and fine aggregates have been removed from the surface so that the coarse aggregates become the major surface feature.

extension line. A line used to indicate the extremities of a feature requiring a dimension.

F

fabrication. Actual work on reinforcing bars or hardware such as cutting, bending, and assembly.

facade. Face or front elevation of a building.

face. The surface of a panel.

face mix. The concrete at the exposed face of a concrete unit; used for specific appearance reasons.

false joint. Scoring on the face of a precast unit; used for aesthetic or weathering purposes and normally made to simulate an actual joint.

fascia. The outside horizontal panel on the edge of a roof or overhang.

fatigue. Weakening due to repeated cycles of stress.

fenestration. The design and placing of windows in a building.

ferrule. The nut-like portion of an insert which receives a machine bolt; it is machined from bar stock.

field. Job site.

filler block. (See end closure.)

FILLET

fillet. A triangular-shaped weld along the interior corner of two steel members that are at right angles.

fin. A projecting vertical nib.

FIN

fineness modulus. An index of fineness or coarseness of an aggregate sample; an empirical factor determined by adding total percentages of an aggregate sample retained on each of a specified series of sieves, and dividing the sum by 100.

fines. Small aggregates such as sand.

finger tight. Tightened until all materials make firm contact; then loosened slightly.

finish. Treatment or texture given to concrete surfaces (see page 166 for various types).

finish mark. A symbol used to indicate the surface to be finished as specified by the mix code or area exposed to view.

fireproofing. Concrete used to surround exposed structural steel members for insulation purposes, thus improving the building's fire resistance.

fire rating. The comparative resistance of a material to failure, as stated in hours, when subjected to standard fire test.

fire-stop. A tight closure of a concealed space with incombustible material to prevent the spreading of a fire.

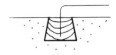

FLANGE

flag. To make note of a change or condition on a drawing.

flange. (1) The horizontal portion of a precast T slab. (2) The projecting edges of a steel beam.

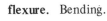

FLASHING REGLET

flashing. Material used to make an exposed intersection weather-tight; materials commonly used are aluminum, sheet metal, and copper.

flashing reglet or slot. A continuous slot cast into a precast panel to receive flashing.

FOOTING

flexure. Bending.

flush. Surfaces in the same plane.

footing. The spread foundation base of a wall or column; generally somewhat wider than the foundation wall.

form. A temporary receptacle which receives concrete and dictates a unit's shape. It can be made of wood or steel, but requires no pattern or positive.

form liner. Molded sheet which when affixed to a form gives the panel a special finish treatment; liners are made of rubber, plastic, wood, etc.

form release agent. A substance applied to the forms for the purpose of preventing bond between the form and the concrete cast in it.

foundation. Building sub-structure supported by earth.

frost line. The distance below grade at which ground will not freeze; locality dictates the exact dimension, so check local codes.

furring. A grid of wooden slats cast into or attached to the back of a panel for the purpose of receiving nailed interior wall construction.

G

gage. A standard scale of measurement for wire diameter and metal thickness. Larger numbers indicate increased thickness (see page 190).

galvanize. To coat with rust-resistant zinc by spraying, dipping, or electro-lytic disposition.

gap-graded concrete. A mix with one or a range of normal aggregate sizes eliminated, and/or with a heavier concentration of certain aggregate sizes over and above standard gradation limits; it is used to obtain a specific exposed aggregate finish.

gap space. The distance between the structure and the back of the panels.

girder. A large, horizontal, structural member used to support the ends of joists and beams or to carry walls over openings.

girt. A horizontal member used to provide wind stability to the structure.

gradation. The sizing of granular materials. For concrete materials, usually expressed in terms of cumulative percentages larger or smaller than each of a series of sieve openings or the percentages between certain ranges of sieve openings.

grade. Reference elevation at top of soil.

grade beam. Low foundation wall or beam usually at ground level, which provides support for the walls of a building.

grade marks. Marking rolled onto a reinforcing bar to identify the grade of steel used in manufacturing.

grade of steel. The means by which a design engineer specifies the strength properties of the steel he requires in each part of a structure, generally using ASTM (American Society for Testing and Materials) designations to distinguish them.

grid lines. Reference lines used on drawings; these lines represent the architectural design grid and often coincide with centerline and/or face of column.

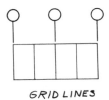

GRID LINES

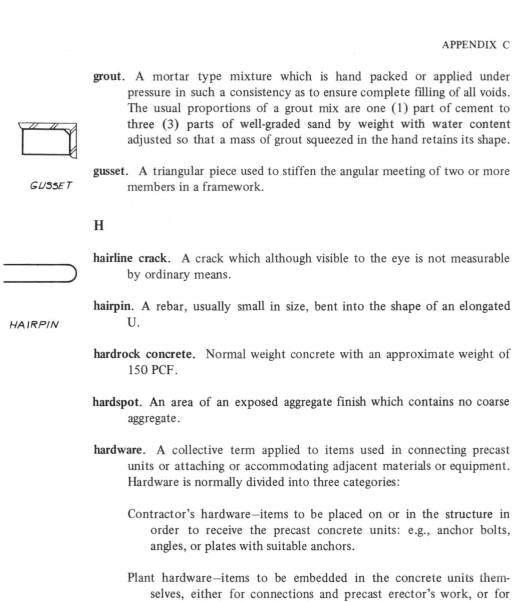

grout. A mortar type mixture which is hand packed or applied under pressure in such a consistency as to ensure complete filling of all voids. The usual proportions of a grout mix are one (1) part of cement to three (3) parts of well-graded sand by weight with water content adjusted so that a mass of grout squeezed in the hand retains its shape.

gusset. A triangular piece used to stiffen the angular meeting of two or more members in a framework.

GUSSET

H

hairline crack. A crack which although visible to the eye is not measurable by ordinary means.

hairpin. A rebar, usually small in size, bent into the shape of an elongated U.

HAIRPIN

hardrock concrete. Normal weight concrete with an approximate weight of 150 PCF.

hardspot. An area of an exposed aggregate finish which contains no coarse aggregate.

hardware. A collective term applied to items used in connecting precast units or attaching or accommodating adjacent materials or equipment. Hardware is normally divided into three categories:

  Contractor's hardware—items to be placed on or in the structure in order to receive the precast concrete units: e.g., anchor bolts, angles, or plates with suitable anchors.

  Plant hardware—items to be embedded in the concrete units themselves, either for connections and precast erector's work, or for other trades, such as mechanical, plumbing, glazing, miscellaneous iron, masonry, or roofing trades.

  Erection hardware—all loose hardware necessary for the installation of the precast concrete units.

haunch. (See corbel.)

HEAD

head. The portion of a panel which forms the top of a window or door opening.

headed stud. Resembling a threadless bolt; its use and installation are the same as a deformed stud.

HEADED STUD

header. A cross member in a series of precast units provided to support one end of the members it interrupts.

hidden lines. Lines behind or beyond the drawn object not seen from the exterior surface.

HEADER

high early. A type of concrete which quickly attains a high compressive strength. This concrete is made with Type III cement.

hollow core. Extruded prestressed slabs varying in thickness from 4 to 12 inches having hollow interior cores to reduce weight.

HOLLOW CORE

honeycomb. A coarse stony concrete surface with voids lacking in fines; some probable causes are congested reinforcement, narrow section, insufficient fines, loss of mortar, and inadequate consolidation.

hook on and/or hook off. The act of placing or removing chokers or slings on or off a member and connecting or disconnecting the crane hook.

hose bib. A hose or faucet connection usually on the exterior of a building at sill height requiring a hole in the precast element.

hydration. The chemical process which takes place between cement and water resulting in hardened concrete.

I

I beam. Rolled steel having a cross-section the shape of an I with a somewhat exaggerated top and bottom horizontal stroke.

I BEAM

impact. A sudden increase in gravity load exerted at stripping or handling which must be considered in handling design.

impact testing. A testing device which measures concrete strength by applying a measurable impact force on the concrete.

insert. A connecting or handling device cast into precast units. Inserts are machine- or coil-threaded to receive a bolt or slotted to receive a bolt head.

insitu. Cast-in-place.

instrument. Transit or level.

interface. The plane between two adjacent surfaces.

J

jacking force. In prestress, the temporary force exerted by the device which introduces the tension into the tendons.

jamb. The vertical sides of a window or door opening.

jig. A template to align parts of an assembly, usually for pre-assembling reinforcing steel cages with a minimum of measurement and consistent accuracy from one cage to the next.

jog. Refers to a shift in the alignment of the edge of a panel.

joint. The space between two adjacent erected panels. (See page 176 for various types.)

joist. A horizontal member in the framing of a floor or roof. These lightweight units are closely spaced and bear on beams or walls.

K

KEY

kerf. To make a cut or notch in a member transversely along the underside in order to curve it; also a cut or notch in a member such as a rustication strip, to avoid damage from swelling of the wood and permit easier removal.

key. A continuous or semicontinuous slot in concrete to receive grout, leveling blocks, or dowels.

key plan. A separate drawing showing a P/C project in plan and used as a common reference for all drawings.

kip. 1,000 pounds.

L

laitance. Residue of weak and nondurable material consisting of cement, aggregate, fines, or impurities brought to the surface of overwet concrete by the bleeding water.

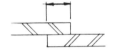

LAP

lally column. The trade name for a cylindrical steel column filled with concrete and used as a vertical support.

lap. The distance two converging members overrun one another.

leader line. Fine lines with an arrowhead touching the edge of the surface referring to a note or dimension.

level. An instrument used for measuring heights of land or other objects above a plane of reference.

lift. The quantity of concrete placed in one operation; or a layer of concrete.

lifting device. An assembly used in handling or erection of precast panels.

LIFT POINT

lifting frame (or beam). A rigging device designed to provide two or more lifting points of a precast concrete element with predictable load distribution and pre-arranged direction of pulling force during lifting.

lift point. Predetermined points from which a panel is to be lifted.

lintel. A horizontal structural member spanning a wall opening at its head to support the wall above the opening.

load bearing. Supporting the dead and live load of other members.

load-bearing precast units. Precast units which form an integral part of the building structure and which are essential to its stability.

loads. (See page 174.)

longitudinal. Situated in the lengthwise direction.

low boy trailer. A trailer with an underslung bed capable of transporting panels of considerable height and still conform to legal height restrictions.

lug. A continuous projection at the sill of a window panel to receive sash.

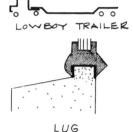

LOWBOY TRAILER

LUG

M

machine thread. A common national coarse thread.

malleability. The ability of a material to undergo considerable plastic deformation under compressive stress.

mansard roof. A roof having slopes on all four sides with the lower slope almost vertical and an upper roof almost horizontal.

MANSARD ROOF

manufacturer's engineer. (See precast engineer.)

marble. A metamorphic rock often irregularly colored by impurities; it is sometimes used as a facing for architectural precast panels.

mark number. The individual identifying mark assigned to each precast unit predetermining its position in the building.

masonry. That branch of construction dealing with the laying up of brick and block with mortar.

masonry opening. Overall opening to be provided in the precast panel.

master mold. A mold which allows a maximum number of casts per project; units cast in such molds need not be identical, provided the changes in the units can be accomplished simply as pre-engineered mold modifications.

mat. A grid of straight reinforcing bars tied together at each intersection.

match line. An imaginary line which separates adjacent areas broken apart for drawing purposes.

material list. Cumulated bills of materials.

matrix. The portion of the concrete mix containing only the cement and fine aggregates (sand).

matte finish. A finish free from gloss or highlights.

maximum size aggregate. Aggregate whose largest particle size is present in sufficient quantity to affect the physical properties of concrete; generally designated by the sieve size on which the maximum amount permitted to be retained is 5 or 10 percent by weight.

mesh. (See welded wire fabric.)

mezzanine. A partial story occurring between two main stories of a building.

MITRE

mitre. The edge of a panel that has been beveled to an angle other than 90°.

mockup. A section of a wall or other assembly built full size, or to scale, for purposes of testing performance, studying construction details, or judging appearance.

module. A repeating or reoccurring dimension or detail.

modulus of elasticity. A measure of the resistance of material to deformation; the ratio of normal stress to corresponding strain for tensile or compressive stresses below the proportional limit of the material; elastic modulus is denoted by the symbol E.

modulus of rupture. Flexural strength of concrete in pounds per square inch (PSI), calculated as the apparent tensile stress in the extreme fiber of a transverse specimen under the load which produces rupture.

mold. The cavity or surface against which fresh concrete is cast to give it a desired shape; sometimes used interchangeably with form but made of fiberglass or concrete; a pattern or positive is built first and the mold is overlayed.

moment. The result of a load on a member, creating a tendency of the member to rotate about a given point or axis that is within its cross section. Moments are measured in foot-kips, foot-pounds, or inch-pounds.

mortar. A mixture of cement, sand, and water; when used in masonry construction, the mixture may contain masonry cement, or portland cement with lime or other admixtures which may produce greater degrees of plasticity and/or durability.

Mo-Sai. Exposed aggregate architectural precast concrete produced under factory-controlled procedures by licensed manufacturers to rigid quality standards established by the Mo-Sai Institute's quality control program.

mosaic. Small colored tile, glass, stone, or similar material arranged to produce a decorative surface.

mud. Plastic concrete (slang).

mullion. A vertical precast unit appearing between windows and/or doors.

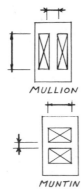

MULLION

MUNTIN

muntin. Horizontal precast appearing between windows and/or doors.

mylar. Fine quality plastic type drafting paper. Requires special pencil leads.

N

nailer. A beveled wooden strip cast into a precast panel for the purpose of nailing flashing or roofing to it. The use of such material in concrete is not recommended.

NAILER

neoprene. A synthetic rubber bearing pad.

neutral axis. Line of zero bending stress of a member when bending occurs in a plane perpendicular to that axis.

nominal. (1) Approximate, not specific; (2) The size of a building material after joints or manufacturing processes are considered.

normal weight concrete. Concrete for which density is not a controlling attribute and usually having unit weights in the range of 135 to 160 lbs. per cubic foot.

nosing. A projection, such as that of the tread of a stair over the riser.

O

object lines. Heavy lines used to indicate the profile of the object drawn. Object lines should be approximately 2½ times as thick as dimension lines.

offset. A displacement or abrupt change in line, or the distance between two parallel lines.

opening. A hole through a panel.

opposite hand. Reverse, mirror image, exact opposite of that shown.

optimum quality. The level of quality, in terms of appearance, strength, and durability, which is appropriate for the specific product, its particular application, and its expected performance requirements. Realistic cost estimates for producing it within stated tolerances are factors which must be considered in determining this level.

overhang. The projecting area of a roof or upper story beyond the wall of the lower part.

P

pachometer. An electronic device used to locate and size reinforcement in hardened concrete.

panel. An individual precast unit.

parapet. That part of a wall that extends above the roof line.

PARAPET

paste. That portion of a concrete mix which comprises 30% by volume; the usual composition of which is 5% air, 15% water, and 10% cement.

patch. To repair a superficially damaged panel by filling the damaged area with concrete of matching color and texture.

pattern or positive. A replica of all or part of the precast element sometimes used for forming the molds in concrete or plastic.

pea gravel. That portion of concrete aggregate passing the 3/8 inch sieve and retained on a No. 4 sieve.

pier. A short column used as a foundation member in building construction.

PIER

pilaster. Column partially or completely embedded in a wall, or a portion of a wall enlarged to serve as a column.

PILASTER

pile. A concrete, steel, or wood member driven into the ground to support a load.

pipe column. A steel cylinder which is used as a vertical support.

plain concrete. Concrete that is either unreinforced or contains less reinforcement than the minimum amount specified for reinforced concrete in ACI 318.

planimeter (compensating polar). An instrument which accurately determines areas of irregular shapes by tracing the perimeter.

plastic concrete. A condition of freshly mixed concrete indicating that it is workable, readily remoldable, cohesive, and has an ample content of fines and cement, but is not over-wet.

plastic cracking. Short cracks often varying in width along their length that occur in the surface of fresh concrete soon after it is placed and while it is still plastic; some probable causes of plastic cracking are high water-cement ratio, low sand content, and poor compaction.

plate. A sheet of metal having a thickness of 1/8 inch or greater.

plate washer. A special washer prefabricated from steel plate.

plug weld. A weld wherein one member partially penetrates another member and the remaining distance is filled with weld.

PLUG WELD

plumb. Vertical; or the act of making vertical.

pole trailer. A trailer used to ship long structural members such as giant tees. This trailer has no bed as such but employs a long pole which fastens the cab to the rear wheels.

post-tensioning. A method of prestressing concrete whereby the tendon is kept from bonding to the concrete, then elongated and anchored directly against the hardened concrete, imparting stresses through end bearing.

poured in place. (See cast-in-place.)

precast concrete. A plain reinforced or prestressed concrete element cast in other than its final position in the structure; precast concrete can be architectural or structural.

precast engineer. The structural engineer authorized by the manufacturer to ensure the adequacy of the structural aspects of the drawings, manufacture, and installation for which the manufacturer is responsible.

preliminary. Not completely finalized and reviewed.

prestressed concrete. Concrete in which there have been introduced internal stresses of such magnitude and distribution that the stresses resulting from loads are counteracted to a desired degree.

prestressing bed. The platform and abutments needed to support the forms and maintain the tendons in a stressed condition during placing and curing of the concrete.

pretensioning. A method of prestressing concrete whereby the tendons are elongated, anchored while the concrete in the member is cast, and released when the concrete is strong enough to receive the stresses from the tendon through bond.

primary dimensions. (1) The basic dimensions of a piece; (2) The first in order of importance.

prime consultant. The architect, engineer, or other professional responsible for the design of the building or structure of which the precast concrete forms a part.

production drawings. (See page 22.)

project drawings. The drawings which accompany project specifications and complete the descriptive information for construction work required or referred to in the project specifications.

P/T conduit. Bright, metallic, flexible, interlocking conduit used in grouted connections.

purlin. A horizontal floor or roof structural member usually resting on joists.

Q

quirk mitre. A corner formed by two chamfered panels (see page 177).

R

rabbet. A two-sided recess frequently used at connections between adjacent units.

rebar. Abbreviated term for reinforcing bar.

reglet. A long, narrow formed slot in concrete to receive flashing or to serve as anchorage.

reinforced concrete. Concrete containing reinforcement, including pre-stressing steel, and designed on the assumption that the two materials act together in resisting forces.

reinforcement. Rebar, mesh, strand, or post-tensioning cables embedded in concrete and located in such a manner that the metal and the concrete act together in resisting loads.

release. (1) The time at which the prestressed strands are severed prior to removing the prestressed units from the forms; (2) submittal of drawings; (3) stripping of precast panel.

release agent. (See form release agent.)

retarder. An admixture which delays the setting of cement paste and therefore of concrete.

retarder, surface. A material used to retard or prevent the hardening of the cement paste on a concrete surface within a time period and to a depth to facilitate removal of this paste after the concrete element is otherwise cured (a method of producing exposed aggregate finish).

retempering. The addition of water and remixing of concrete which has started to stiffen in order to make it more workable.

RETURN

return. A projection of like cross-section which is 90° to or splayed from main face or plane of view.

REVEAL

reveal. (1) groove in a panel face generally used to create a desired architectural effect; (2) The projection of the coarse aggregate from the matrix after exposure.

revibration. Delayed vibration of concrete that has already been placed and consolidated; most effective when done at the latest time a running vibrator will sink of its own weight into the concrete and again make it plastic.

RIB

rib. (1) Continuous vertical projection on a wall panel projecting a minimum of 6 inches from the panel face; (2) Local thickening providing stiffness in concrete panels.

rigger. Mechanic whose function is to brace, guy, and arrange for hoisting materials.

rod. Term used to describe any of a number of types of round steel bars.

rolled section. Structural steel member, such as an I beam or wide-flange section, that is formed into its shape by hot rolling at the mill.

rolling block. Pulley used during panel handling to evenly distribute loads to inserts or to facilitate rotating the panel into its final position.

ROLLING BLOCK

rustication. A groove in a panel face for architectural appearance; also reveal.

rustication strip. A strip of wood or other material attached to a form surface to produce a groove or rustication in the concrete.

S

sacking. A common remedy for pits and air bubble holes in concrete—a slurry (the consistency of thick cream) consisting of a mixture of sand and cement is thoroughly rubbed over the moistened area with clean burlap pads or sponge rubber floats.

safety factor. Number that results from dividing the ultimate strength by the allowable working stress allowing for any imperfections or additional loads; codes regulate the minimum safety factor required in many areas.

safe working load. That magnitude of load which a connection can safely resist while maintaining factors of safety.

sand. That portion of an aggregate passing the No. 4 (4.76 mm) sieve and predominantly retained on the No. 200 (74 micron) sieve.

sandblast. A system of abrading a surface such as concrete by a stream of sand, or other abrasive, ejected from a nozzle at high speed, by water and/or compressed air.

sandwich wall panel. Panel consisting of two layers (wythes) of concrete fully or partly separated by a layer of insulation; in employing this detail, metal shear connectors are usually required to tie the two layers of concrete together.

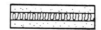

SANDWICH PANEL

scabbing. A finish defect in which parts of the form face including release agent adhere to the concrete, some probable causes are an excessively rough form face, inadequate application of release agent, or delayed stripping.

scaling. A finish defect resulting in a thin layer of hardened mortar breaking free from the concrete surface and exposing mortar or stone; some probable causes are low strength concrete, rough or absorbent form face, inadequate application of release agent and curing procedure.

schokbeton. Shocked concrete; a method of precasting architectural concrete in which the mold is fastened to a steel shocking table and repetitively raised a fraction of an inch and dropped at the rate of about 250 impacts a minute to consolidate the concrete.

score. To use a saw to notch hardened concrete.

scouring. Irregular eroded areas or channels with exposed stone or sand particles; some probable causes of this finish defect are excessively wet concrete mix, insufficient fines, water in form when placing, poor vibration practices, and low temperature when placing.

screed. A wooden or metal tool used to level off the back surface of a panel flush with the form side rails.

scupper. An opening in a wall panel through which the roof is drained.

SCUPPER

sealants. A group of materials used to seal joints between precast concrete units and between such units and adjacent materials.

sealers or protective coatings. Materials used to coat precast concrete units for the purpose of improving resistance to water penetration or for improving weathering qualities.

secondary dimensions. Related directly to the primary dimensions and of secondary rank or value. They must always total the primary dimensions.

section. Cut away view through a general plan or elevation view to explain details.

section indicator. A symbol indicating the direction in which a cross-section is cut or viewed.

section mark. A letter or number on a drawing indicating the location of a section on the drawing.

segmental member. Structural member made up of individual elements prestressed together to act as a monolithic unit under service loads.

segregation. The tendency for the coarse particles to separate from the finer particles in handling; in concrete, the coarse aggregate and drier material remains behind and the mortar and wetter material flows ahead; this also occurs in a vertical direction when wet concrete is overvibrated or dropped vertically into the forms, the mortar and wetter material rising to the top; in aggregate, the coarse particles roll to the outside edges of the stockpile.

selfstressing forms. Equipment which in addition to serving as forms for concrete also accommodates the pre-tensioned strands (or wires) and sustains the total prestressing force by suitable end bulkheads and sufficient cross-sectional strength.

sepia. A brown on white, transparent, reproducible print.

set. Erect, place, install.

set plate. A steel plate which is pre-installed and grouted to the desired elevation and line for the purpose of receiving a precast column or panel.

set-up. The process of preparing molds or forms for casting, including locating materials (reinforcement and hardware) prior to the actual placing of concrete.

shackle. (See swivel plate.)

shear. The result of two parallel forces acting on a body in opposite directions tending to cause two parts of the body to slide against each other.

shear key. A continuous slot formed expressly to receive plastic mortar or concrete to resist lateral separation.

SHEAR KEY

shear wall. Wall designed to resist forces resulting from wind, blast, or earthquake.

shim. Material placed between a panel or its connecting angle and the supporting structure for the purpose of controlling the panel's vertical alignment, usually steel.

SHIM

shim space. The space between floor (or beam) levels and bearing area of precast connections.

shop drawings. (See page 20.)

shrinkage. The volume change in precast units normally occurring during the hardening process of concrete.

side rail. The side of a form.

sill. That portion of a precast panel which forms the bottom or base of a window or door opening.

SILL

slab. Precast or prestressed floor or roof members.

sleeve. Any cylinder cast into a panel for the purpose of creating a hole.

sling. Short lengths of wire rope, with a spliced eye at each end or a spliced eye at one end and a hook at the other, used in erection of small precast units; use of a sling eliminates the necessity of casting erection inserts into a panel.

slot. A hole having a length of approximately 2½ times its diameter.

slug. A short length of rebar used as a filler in completing welds.

SLUG

slump. Measure of the consistency of plastic concrete, equal to the number of inches of subsidence of a truncated cone of concrete released immediately after molding in a standard slump cone.

slump cone. Metal mold in the form of a truncated cone with a top diameter of 4 inches, a base diameter of 8 inches, and a height of 12 inches, used to fabricate the specimen for the slump test.

SOFFITT

slurry. Thin mixture of water and finely divided materials, such as portland cement, in suspension.

soffitt. A precast panel used to form the underside of an overhang or ceiling; also, the finished underside of a lintel beam, or cantilevered floor.

spalling. A finish defect, more severe than scaling, in which pieces of concrete break free from the hardened surface; some probable causes are low concrete strength, absorbent aggregates which are susceptible to damage by frost or water, inadequate drafts in form work, inadequate release agent, early stripping, and corrosion of reinforcement.

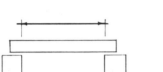

SPAN

span. The horizontal distance between supports of a member such as a beam, girder, slab, or joist.

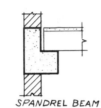

SPANDREL BEAM

spandrel. That part of a wall between the head of a window and the sill of the window above it. (An upturned spandrel continues above the roof or floor line.)

spandrel beam. Beam in a building frame which extends between exterior columns at a floor level.

specifications. The type or printed directions issued by architects to establish general conditions, standards, and detailed instructions which are used with the contract drawings; contracted term, **specs.**

spiral reinforcement. Reinforcing rod or wire supplied in coils and used in place of ties for columns.

splay. A beveled or slanted surface. (See draft.)

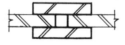

SPLICE

splice. Connection of one reinforcing bar to another by lapping, welding, mechanical couplings, or other means; the lap between sheets or rolls of welded wire fabric.

spot weld. A quickly applied weld which is not required to perform structurally.

SPREADER BEAM

spreader beam. A frame of steel channels or beams attached to the back of a panel, prior to stripping, for the purpose of evenly distributing loads to inserts and for lifting the panel about its center of gravity.

STEM

stainless steel. Steel alloyed with sufficient chromium to resist corrosion, oxidation, or rusting.

stem. The vertical leg of a prestressed tee.

stiffener. A steel plate welded to a steel beam or column to increase the section's stiffness at a desired point.

STIFFENER

stirrup. A rebar bent into the form of a closed tie or U-shape and surrounding the main reinforcement in panels and beams for control of cracking caused by diagonal tension; the term **stirrups** is usually applied to lateral reinforcement in horizontal members, and the term **ties** to those in vertical members.

STIRRUP

stone anchor. An anchor commonly used to fasten cut stone units; such anchors are seldom suitable for use in precast concrete attachment.

strain. The measure of a member's deformation.

strand. Tendon usually composed of three- or seven- wire assemblies used as reinforcement in prestressed concrete.

strand chuck or vice. A device for holding a strand under tension.

stress. Intensity of force per unit area.

stringer. (1) The supporting member of a stair upon which the treads are laid; (2) horizontal structural member usually (in slab forming) supporting joists and resting on vertical supports.

STRINGER

stripping. The process of removing a precast concrete element from the form in which it was cast.

strong-back. A steel or wooden plate which is attached to a panel for the purpose of adding stiffness during handling, shipping, and/or erection.

structural. A unit which carries live load or another unit's weight.

sub-contractor (sub). A contractor who provides materials or services to the general contractor.

submitted. Presented to the architect/engineer for review.

superstructure. That portion of any building which extends above the foundation.

swivel plate. A clevis which can be bolted to a panel, thus providing a hook for use in handling.

symmetry. Exact correspondence of shape on opposite sides of a dividing line or plane.

systems building. Essentially the orderly combination of "parts" into an "entity" such as sub-systems or the entire building; systems building makes full use of industrialized production, transportation, and assembly.

T

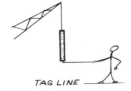

TAG LINE

tack weld. (See spot weld.)

tag line. A rope attached to a precast unit during erection for use in helping the crane operator guide the panel into place. It is manipulated by hand from ground level.

tamp. To pack concrete down tightly by a succession of blows or taps.

TEE

tee. A structural floor or roof member consisting of one or more stems and a thin flange. Tees are sometimes used as architectural wall panel.

teflon pad. A bearing pad made of a waxy opaque material. These pads are used when restraint of movements is not desired.

temperature reinforcement. Reinforcement distributed throughout the concrete to minimize cracks due to temperature changes and concrete shrinkage.

temperature rise. The increase of concrete temperature caused by heat of hydration and heat from other sources.

template. A pattern made of thin metal plate or plastic and used as a guide in accurately positioning reinforcement anchor bolts or inserts; double templates are often used to help ensure plumb.

tendon. Tensioned element, generally high-strength steel wires, strands, or bars, used to impart prestress to the concrete; in post-tensioned concrete, the complete assembly of prestressing steel, anchorages, and sheathing, when required, is also called a tendon.

tendon (bonded). Tendon which is bonded to the concrete through grouting or other approved means, and therefore is not free to move relative to the concrete.

tendon (unbonded). Tendon in which the prestressing steel is permanently free to move relative to the concrete to which it is applying the prestressing forces.

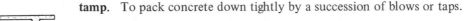

TENSION

tension. Stress or force in a material caused by a pulling action, which tends to create a lengthening of the material.

terrazzo. Flooring surface of marble chips in concrete which is ground and polished after setting.

texture. Any finish other than a smooth finish.

thermal movements. Volume changes in precast units caused by temperature variations.

threaded. Having national coarse machine thread.

tie. A closed loop of small size reinforcing bars that encircle longitudinal bars in columns and beams. (See stirrup.)

tie wire. (See bagtie.)

tilt-up. Method of concrete construction where members are cast horizontally near their eventual position, and then tilted into place after removal of forms.

tolerance. Specified permissible variation from stated requirements such as dimensions, strength, and air-entrainment.

TOOL

tool. To slightly round a corner, in freshly cast concrete with a cement mason's edger.

tooling. Most of the manufacturing and service processes preceding the actual set-up and casting operations.

topping. Concrete cast on erected prestressed units to achieve a level floor or to aid the units in uniformly carrying loads.

torsion. The stress caused when one portion of a member is twisted in one direction and the other end is held motionless or twisted in the opposite direction.

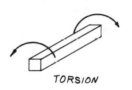

TORSION

tracing. A reproducible original drawing, white background with black printing.

transit. A surveying instrument used to measure horizontal and vertical angles.

transverse. At right angles to the long direction of the member (crosswise); also referred to as **lateral**.

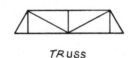

TRUSS

truss. Structural members arranged and fastened in triangular units to form a rigid framework for support of loads over a long span.

T-shore. An inverted tee-shaped concrete unit used to support precast panels during storage.

T-SHORE

turning table. Mechanical table used in precast plants to rotate units from the horizontal casting position into the vertical handling position without the necessity of handling inserts; when employed the amount of rebar in a panel is greatly reduced.

U

UL approved. Rated and approved for performance during a fire by the Underwriter's Laboratories, Inc.—a non-profit organization; ratings are in units of hours based on tests.

ultimate strength. Maximum resistance to loads that a structure or member is capable of developing before failure occurs, or, with reference to cross-sections of members, the largest axial force, shear, or moment a structural concrete cross-section will support.

V

vacuum pad. Hydraulic suction cups used for handling flat panels; a smooth finish is required if such a device is to be used.

vellum. A heavy off-white translucent fine quality tracing paper.

veneer. A layer of facing material, such as natural stone, used to cover a panel.

vibration. Energetic agitation of concrete to assist in its consolidation, produced by mechanical oscillating devices at moderately high frequencies; external vibration employs a device attached to the forms and is particularly applicable to the manufacture of precast items; internal vibration employs an element which can be inserted into the concrete, and is more generally used for cast-in-place construction.

W

wall (bearing). A wall supporting a vertical load in addition to its own weight.

wall panel. A component of a prefabricated wall which derives its strength and dimensional stability from a precast concrete element; the component includes any nonconcrete items incorporated in the element at the time of manufacture.

WASH

warping. The bowing of a precast unit in two planes.

wash. The sloped surface of a sill panel which permits rainwater to run off.

waterproofing. An application which when sprayed onto a panel surface tends to repel rainwater.

weatherproofing. The process of protecting all joints and openings from the penetration of moisture and wind.

WEB

weather sealing. The process of treating wall areas for improved weathering properties.

web. That portion of a beam or precast unit to which the flanges are attached.

wedge insert. An insert having a wedge shaped holding face which permits vertical adjustment without slippage through the use of a special askew head bolt (see page 180).

weep hole. A hole provided for drainage through precast panel joints.

weld. To join metals by applying heat with a filler metal which has a high melting point.

welded wire fabric. A reinforcing material composed of cold drawn steel wires fabricated into a sheet consisting of longitudinal and transverse wires arranged at right angles and welded together at all points of intersection.

weld plate. A plate with attached anchors cast into concrete for the purpose of making a welded connection.

WELD PLATE

wet-mix concrete. Concrete mixtures designed for typical water-cement ratios, slumps, and handling and consolidation methods.

whiteprint. A nonreproducible print, made from a tracing, and having white background with blue or black printing.

wide flange. An H-shaped rolled steel section.

WIDE FLANGE

winch. Mechanical lifting device attached to derricks on which cable is wound up by means of a crank and locked in position by a ratchet.

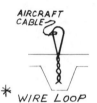

AIRCRAFT CABLE

wire loop. A lifting loop cast into the back of a precast member for handling purposes; this loop is usually formed with discarded prestress strand or aircraft cable.

* WIRE LOOP

workability. The ease with which a given set of materials can be mixed into concrete and subsequently handled, transported, placed, and finished with a minimum loss of homogeneity.

working point. A real point from which a dimension (panel) is measured. It must be accessible to workmen.

working stress. The maximum unit stress considered desirable in a structural member subjected to loads.

work point. A real, imaginary, or inaccessible point from which a dimension on a drawing is measured.

wrinkled tin. Slang for corrugated metal.

wythe. A continuous vertical section of a wall tied to its adjacent vertical element (part of a composite wall).

Z

zipper gasket. A neoprene rubber window gasket used to fasten window glass to a precast panel.

ZIPPER GASKET

Some draftsmen insist upon abbreviating every word possible on their drawings, mainly to economize; other draftsmen, knowing the inconsistencies and chances of misinterpretation, will not tolerate the use of abbreviations. The draftsman must use discretion, however, and abbreviate only those words that have commonly identified abbreviations and use them in such a way that they cannot possibly be misconstrued. Capital lettering is usually employed, and as a rule, the period at the end is omitted. Abbreviations common to precast concrete are listed in this section. All abbreviations used on the drawings, should be defined on the cover sheet.

A

addendum number	ADD NO
additional	ADTNL
adjacent	ADJ
adjustable	ADJT
after weld	AW
aggregate	AGG
air condition	A/C
alternate	ALT
aluminum	AL
American Concrete Institute	ACI
American Institute of Architects	AIA
American Institute of Steel Construction	AISC
American National Standards Institute	ANSI
American Society of Civil Engineers	ASCE
American Society for Testing and Materials	ASTM
American Welding Society	AWS
American wire gage	AWG
amount	AMT
anchor	ANC
anchor bolt	AB
anchor plan	AP
angle	L or (∠)*
approved	APPD
approximate	APPROX
architect	ARCH
area	A
asbestos	ASB
assembly	ASSY
associates	ASSOC
association	ASSN
auxiliary	AUX
avenue	AVE

B

back to back	B/B
base plate	BP
baseline	BL
basement	BSMT
beam	BM
bearing	BRG
bench mark	BM
bent	BT
between	BET
bill of material	BOM
block	BLK
blocking	BLKG
board	BD
both faces	BF
both sides	BS
bottom	BOT
brick	BRK
building	BLDG
bulkhead	BHD
by others	BO

C

cadmium	CAD
calculation	CALC
cantilever	CANT
capacity	CAP
cast-in-place	CIP
catalog	CAT
caulking	CLKG
ceiling	CLG
cement	CEM

*Use symbol

cement asbestos	CA	durometer	DURO
center	CTR		
center to center	C/C	*E*	
center of gravity	CG		
centerline	(₵)*	each	EA
chamfer	CHAM	each face	EF
channel	(c)*	each way	EW
check	CK	east	E
checked	CKD	electrical	ELEC
chord	C	elevation	EL
circumference	CIRC	elevator	ELEV
civil engineer	CE	end to end	E/E
clear	CLR	engineer	ENGR
column	COL	engineering	ENGRG
company	CO	equal	EQ
composite	COMP	erection	ERECT
compressive strength at stripping	f'_{ci}	existing	EXIST
compressive strength at 28 days	f'_c	expansion	EXPAN
concrete	CONC	expansion joint	EJ
concrete masonry unit	CMU	exposed	EXP
Concrete Reinforcing Steel Institute	CRSI	exposed aggregate concrete	EAC
connection	CONN	exterior	EXT
construction	CONST		
Construction Specifications Institute	CSI	*F*	
continued	CONTD		
continuation	CONTIN	fabricator	FAB
control joint	CJT	face to face	F/F
continuous	CONT	far face	FF
contractor	CONTR	Federal Housing Administration	FHA
corner	COR	figure	FIG
corporation	CORP	finish	FIN
counterflashing	CTFL	finished	FD
countersink	CS	fireproof	FP
cubic	CU	flashing	FLG
		floor	FLR
		foot	FT
		footing	FTG
D		foundation	FDTN
dead load	DL	*G*	
deflection	DEFL		
deformed anchor stud	DAS	gallon	GAL
department	DEPT	galvanize	GALV
detail	DTL	gauge or gage	GA
diagonal	DIAG	general	GEN
diameter	DIA or (ϕ)	general contractor	GC
dimension	DIM	girder	G
distance	DIST	glass	GL
ditto	DO	government	GOVT
double tee	DT	grade	GR
dovetail anchor	DTA		
dovetail slot	DTS	*H*	
down	DN		
drafting	DFTG		
draftsman	DFTMN		
drain	DR		
drawing	DWG	half round	HR

*Use symbol

hardrock	HDRK	longitudinal	LONGT
hardware	HDWR	long leg horizontal	LLH
headed anchor stud	HAS	long leg vertical	LLV
height	HT	low point	LP
hexagonal	HEX		
high point	HP	*M*	
high strength	HS		
high strength bolt	HSB	machine	MACH
hollow metal	HM	machine bolt	MB
horizontal	HORIZ	malleable iron	MI
hour	HR	manufacturer	MFGR
		manufacturing	MFG
I		mark	MK
		masonry	MAS
inch	IN	masonry opening	MO
include	INCL	match line	ML
incorporated	INC	material	MTL
information	INFO	maximum	MAX
insert	INS	mechanical	MECH
inside diameter	ID	metal	MET
insulation	INS	mezzanine	MEZZ
interior	INT	middle	MID
		minimum	MIN
J		minute	(')*
		miscellaneous	MISC
joint	JT	mixture	MIX
joints	JTS	modification	MOD
		modular	MR
K		moment of inertia	I
		mullion	MULL
kip	K		
kip per lineal foot	KLF	*N*	
kip per square inch	KSI		
kip per square foot	KSF	National Bureau of Standards	NBS
knockout	KO	national coarse	NC
		national fine	NAF
L		near face	NF
		negative	NEG
laboratory	LAB	neutral axis	NA
laminated	LAM	normal weight concrete	NWC
left	L	north	N
left hand	LH	not in contract	NIC
left side	LS	not to scale	NTS
length	L	number	NO or (#)
length X 0.207 (fifth point)	L/5		
light	LT	*O*	
light weight	LWT		
light weight concrete	LWC	on center	OC
limited	LTD	opening	OPG
lineal	LIN	opposite	OPP
live load	LL	optional	OPT
location	LOC	original	ORIG
long	LG	out to out	O/O

*Use symbol

outside diameter	OD
overall	OA
overhand	OHG
overhead	OH

P

pair	PR
panel	PNL
parallel	PAR
partition	PTN
perpendicular	PERP
piece	PC
plate	PL or (℄)*
plumbing	PLMB
point	PT
polyvinyl chloride	PVC
porcelain	PORC
Portland Cement Association	PCA
post tensioned	P/T
pound	LB or (#)
pounds per cubic foot	PCF
pounds per lineal foot	PLF
pounds per square foot	PSF
pounds per square inch	PSI
precast concrete	P/C
preliminary	PRELIM
premolded joint filler	PJF
prestressed	P/S
Prestressed Concrete Institute	PCI
production	PROD
project	PROJ
projection	PROJN

Q

quality	QUAL
quantity	QTY
quarter round	QR

R

radius	R
received	RECD
recessed	RESD
rectangular	RECT
reference	REF
reinforcement	REINF
required	REQD
revision	REV
right	RT
right hand	RH
right side	RS
road	RD
roof	RF
room	RM
round	RND or (ϕ)

S

safe working load	SWL
same as	SA
schedule	SCH
section	SECT
section modulus	S
sheet	SHT
similar	SIM
sketch	SK
south	S
spaces or spacing	SPA
specifications	SPEC
square	SQ
staggered	STGR
stainless steel	SST
standard	STD
standard wire gage	SWG
steel	STL
stirrup	STIR
street	ST
structural	STR
sub-contractor	SUB
superimposed	SUPER
symmetrical	SYM
system	SYS

T

that is	IE
thickness	T
thousand	M
thread or threaded	THRD
tongue and groove	T & G
top and bottom	T & B
top of concrete	T/C
top of precast	T/P
top of steel	T/S
total	TOT
transverse	TRANS
typical	TYP

U

ultimate	ULT
Underwriter's Laboratories	UL
uniform building code	UBC
United States gage	USG

*Use symbol

unless noted otherwise	UNO	wide flange	(W,M,S)*
		width	B
V		width X 0.207 (fifth point)	B/5
		window	WDO
versus	VS	Wire Reinforcement Institute	WRI
vertical	VERT	with	W/
volume	VOL	without	W/O
		wood	WD
W		work point	WP
waterproofing	WPFG	*Y*	
weight	WT		
welded wire fabric	WWF	yard	YD
west	W		

*Use symbol

The importance of tolerances increases with each advanced step in the industrialization of the building process. Draftsmen must be familiar with this subject in order to utilize these advances and to properly dimension the precast units. Three groups of tolerances which should be established as part of architectural precast concrete design and to which final unit details should conform are: tolerances for erection; tolerances for manufacturing; and tolerances for interfacing. The PCI *Architectural Precast Concrete Manual* should be consulted for a discussion on industry tolerances.

Tolerances for Erection

Definition. Erection tolerances are those required for realistic matching with the building structure and will normally involve the General Contractor and/or different sub-contractors.

Example. Clearance between the structural building frame and the precast concrete wall. Also the location of connection elements for precast panels (Contractor's hardware).

Selection. The basis for erection tolerance is determined by the characteristics of the building structure and site conditions.

Erection tolerances related to precast panel work are listed in Table E-1. Normally, cast-in-place concrete and structural steel have far less restrictive tolerances than a precast concrete panel. Tolerances for steel are taken from the *Manual of Steel Construction*, and those for concrete from ACI 347, *Recommended Practice for Concrete Formwork*. Tolerances for cast-in-place structures may have to be increased further to reflect local trade practices, the complexity of the structure, and climatic conditions. The stated cast-in-place tolerances are difficult to attain during severe winter conditions. When detailing precast units for attachment to steel structures, allowance must be made for sway in tall, slender steel structures with uneven loading, and movements caused by daily temperature changes.

Ample distance should be allowed between the theoretical face of the building and the back face of the panels. A minimum of 1 inch is required and at least 1½-2 inches in tall, irregular structures. If this primary distance is realistically assessed, it will solve many tolerance problems. Where large tolerances have been allowed for a supporting structure—or where no tolerances are given or appear enforceable—this dimension must be increased.

In the determination of tolerances, attention should also be given to possible deflections and/or rotation of structural members supporting precast concrete. This is particularly important for bearing on slender or cantilevered structural members. Consideration should be given to both initial deflection and to long-term deflections caused by plastic flow (creep) of the supporting structural members.

155

Table E-1 INTERFACING ERECTION TOLERANCES SHOWING RELEVANT ERECTION TOLERANCES FOR CAST-IN-PLACE CONCRETE STRUCTURES AND ERECTION TOLERANCES FOR STRUCTURAL STEEL

VARIATION OR TOLERANCE	CAST-IN-PLACE CONCRETE	STEEL
Variations from the plumb, or column tolerances	¼ inch per 10 feet but not more than 1 inch Valid to 100 feet height. No tolerances suggested above 100 feet	1 to 1000, no more than 1″ towards building nor 2″ away from building line in the first 20 stories; plus 1/16″ for each additional story up to a maximum of 2″ towards building or 3″ away from building line
Tolerances in levels	In 10 feet.................... ¼″ Up to 20 ft. bay.............. ⅜″ In 40 ft. or more .... ¾″	Erection tolerances for levels normally not stated, as levels should be governed by close manufacturing tolerances
Variations from the linear building lines in relation to columns and walls	In any bay................... ½″ In bay 20 ft. max............. ½″ In bay 40 ft. max............. 1″	As set by column alignments Closer for elevator columns
Tolerances in beams and columns	Cross section dimensions $\rbrack$ —¼″ +½″	1 to 1000 in alignment Section tolerances are close
Tolerances for placing or fastening of other materials such as connection hardware in relation to building lines	Not established	Not established

Final erection tolerances should be verified and agreed on when erection commences and, if different from those originally planned, stated in writing or noted on erection drawings.

Tolerances for Manufacturing

Definition. Manufacturing tolerances are those inherent in any manufacturing process. They are normally determined by economic consideration and functional and appearance requirements. The actual tolerances should, however, be related to the amount of repetition and the size and other characteristics of the precast unit.

Example. Physical dimensions of panels such as length, width, squareness, and openings.

Selection. Tolerances can be made smaller than for erection due to plant-controllable operations.

Tolerances for manufacturing are becoming standardized throughout the industry and should only be made more severe, and therefore more costly, where absolutely necessary. These areas might be special finish or appearance requirements, some glazing details, and certain critical dimensions of open-shaped panels.

Suggested tolerances for manufacture and installation of architectural precast concrete panels* are as follows:

1. Warpage and bowing. Wall panels shall be manufactured and installed so that each panel after erection complies with the following dimensional requirements:

> Maximum permissible warpage of one corner out of the plane of the other three shall be the greater of 1/16 inch/foot distance from the nearest adjacent corner or 1/8 inch.

This requirement is illustrated in figure E-1.

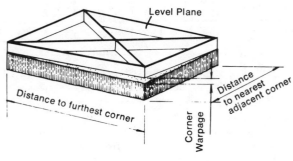

Figure E-1

> Bowing—concave or convex—of any part of a flat surface shall not exceed

$$\frac{\text{length of bow}}{360}$$

with a maximum of 3/4 inch up to 30 feet. This requirement is illustrated in figure E-2.

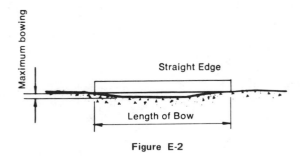

Figure E-2

*For precast units, such as columns, beams, etc. tolerance requirements are given in PCI MNL-116, *Manual for Quality Control of Plants and Production of Precast Prestressed Concrete Products.*

Slender panels cannot automatically be subjected to the standard tolerances for bowing and warping. Table E-2 gives width and length dimensions in relation to panel thicknesses, below which warping tolerances should be reviewed for each individual project. For slender panels below the range shown in Table E-2 and/or for panels with large aggregates, above 3/4 inch, these tolerances should be increased. Tolerances for units which are not homogeneous—consisting either of two widely different concrete mixes or natural stone veneer with concrete back-up—should be specifically reviewed and may also have to be increased.

Table E-2 CHART OF PANEL THICKNESSES

Panel Dimensions	8'	10'	12'	16'	20'	24'	28'	32'
4'	3"	4"	4"	5"	5"	6"	6"	7"
6'	3"	4"	4"	5"	6"	6"	6"	7"
8'	4"	5"	5"	6"	6"	7"	7"	8"
10'	5"	5"	6"	6"	7"	7"	8"	8"

> **This table represents a relationship between overall flat panel dimensions and thicknesses below which suggested warpage tolerances should be reviewed and possibly increased. For ribbed panels the equivalent thickness should be the overall thickness of such ribs if continuous from one end of the panel to the other.**

2. Dimensional tolerances of finished units. Overall height and width measured at the face adjacent to the mold at time of casting:

10 feet or under	+ 1/8 inch
10 feet to 20 feet	+ 1/8 inch, – 3/16 inch
20 feet to 30 feet	+ 1/8 inch, – 1/4 inch
Each additional 10 feet	+ 1/16 inch per 10 feet
Angular deviation of plane of side mold	1/32 inch per 3 inch depth or 1/16 inch total, whichever is greater
Openings within one unit	+ 1/4 inch
Out of square (difference in length of the two diagonal measurements)	1/8 inch per 6 feet or 1/4 inch total, whichever is greater

Thickness: Total thickness shall be within – 1/8 inch + 1/4 inch

Tolerances on any dimension not specified above: The numerically greater of ± 1/16 inch per 10 feet or + 1/8 inch.

The location of these tolerances for a typical
wall panel is shown in figure E-3.

WALL PANEL TOLERANCES

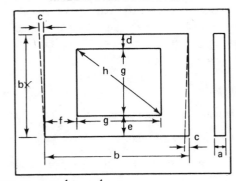

a. Thickness ($-\frac{1}{8}$ in. $+\frac{1}{4}$ in.)
b. Height and width
       10 ft. and under $\pm \frac{1}{8}$ in.
       10 ft.-20 ft. $+\frac{1}{8}$ in.-$\frac{3}{16}$ in.
       20 ft.-30 ft. $+\frac{1}{8}$ in.-$\frac{1}{4}$ in.
       Each additional 10 ft. $\pm \frac{1}{16}$ in.
c. Out of square: $\frac{1}{8}$ in. per 6 ft. but not greater than $\frac{1}{4}$ in.
d. Opening location from top: $\pm \frac{1}{8}$ in.
e. Opening location from bottom: $\pm \frac{1}{8}$ in.
f. Opening location from side: $\pm \frac{1}{8}$ in.
g. Length and width of opening: $\pm \frac{1}{4}$ in.
h. Squareness of opening (diagonal): $\pm \frac{1}{4}$ in.
i. Bowing:
    $\frac{\text{Length of bow}}{360}$ with $\frac{3}{4}$ in. max.
                                         up to 30 ft. (360 in.)
    Warpage: $\frac{1}{16}$ in. per foot from nearest adjacent corner or $\frac{1}{8}$ in.

Figure E-3

3. Position tolerances. For cast-in-items measured from datum
line locations as shown on the approved erection and production
drawings:

Anchors and inserts shall be within 3/8 inch of centerline location
as shown on drawings.

Blockouts and reinforcement shall be within 1/4 inch of the
position as shown on the drawings, where such positions have
structural implications or affect concrete cover; otherwise they
shall be within ± 1/2 inch.

All reinforcing steel should have a minimum cover of 3/4 inch
±1/4 inch and shall be accurately located as indicated on the
approved shop drawings. The 3/4 inch cover is realistic only if the
maximum aggregate size does not exceed 1/2 inch. Cover
requirements over reinforcement should be increased when the
precast units are exposed to a corrosive environment or severe
exposure conditions. For exposed aggregate surfaces, the 3/4 inch
cover to surface of steel should not be measured from the original
surface; the typical depth of mortar removal between the pieces
of coarse aggregate should be subtracted.

Where possible, connections should be sized to the nearest 1/2
inch. This makes it easier to detail connections and it simplifies
production. Also, 1/2 inch increments are common in plate sizes
up to 6 inches.

Coordination of the dimensions of the elements within connections leads to satisfactory production of the connection detail. Dimensional considerations require reasonable clearances and tolerances. It is not practical nor economical to have the various connection elements assembled like a watch. Therefore, the minimum clearance between various items within a connection should not be less than 1/4 inch, with 1/2 inch preferred. (See figure E-6 for detailing of connections.)

Flashing reglets	± 1/4 inch
Flashing reglets, at edge of panel	± 1/8 inch
Reglets for glazing gaskets	± 1/16 inch
Groove width for glazing gaskets	± 1/16
Electrical outlets, hose bibs, etc.	± 1/2 inch

Location of hardware items cast into, or fastened to the structure by the General Contractor or other trades should be determined with specified tolerances for placement. Average tolerances for such locating dimensions should be ± 3/4 inch in all directions, plus a slope deviation of no more than ± 1/8 inch for the level of critical bearing surfaces. For smaller and simpler structures, location tolerances of +1/2 inch may be feasible. Complicated structures should allow 1 inch for accuracy of location.

4. Tolerances for location of precast units: Precast units shall be manufactured so that joints between the panels are within these limits:

Face width of joints	± 3/16 inch
Joint taper	1/40 inch per foot length, with a maximum length of tapering in one direction of 10 feet
Step in face (Fig. E-4).	1/4 inch
Jog in alignment of edge (Fig. E-5).	1/4 inch

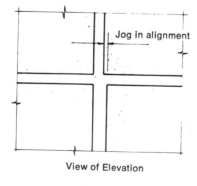

View of Elevation

Figure E-4

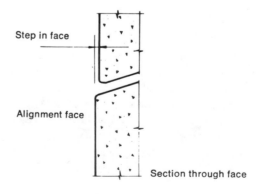

Section through face

Figure E-5

DETAILING CONNECTIONS FOR TOLERANCE (CURTAIN WALL)

Types of tolerance:
A. Initial — to compensate for normal precasting and construction errors
B. Long term — to allow panel to undergo volume changes without damage

Directions of Tolerances

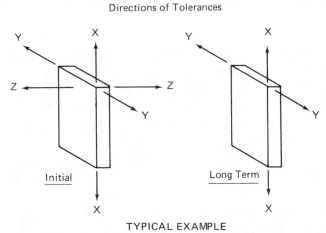

Initial

Long Term

Initial X (vertical) required to compensate for misplacement of anchors in precast and to allow for vertical misalignment of structure.
Initial Y (horizontal) required to compensate for misplacement of anchors in precast and in structure (if concrete)
Initial Z (in and out) required to compensate for in and out variation of structure and variation in precast thickness.
Long term X (vertical) to accommodate volume changes in precast and/or deflections or cambers in the structure.
Long term Y (horizontal) to accommodate volume changes in precast.

TYPICAL EXAMPLE

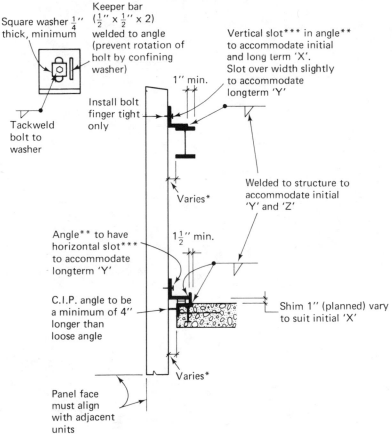

Square washer $\frac{1}{4}''$ thick, minimum

Keeper bar ($\frac{1}{2}''$ x $\frac{1}{2}''$ x 2) welded to angle (prevent rotation of bolt by confining washer)

Tackweld bolt to washer

Install bolt finger tight only

1" min.

Vertical slot*** in angle** to accommodate initial and long term 'X'. Slot over width slightly to accommodate longterm 'Y'

Varies*

Welded to structure to accommodate initial 'Y' and 'Z'

Angle** to have horizontal slot*** to accommodate longterm 'Y'

$1\frac{1}{2}''$ min.

C.I.P. angle to be a minimum of 4" longer than loose angle

Shim 1" (planned) vary to suit initial 'X'

Panel face must align with adjacent units

Varies*

*** Slots to be $2\frac{1}{2}$ times bolt diameter
** $\frac{3}{8}''$ min. angle thickness for welding
* Construction tolerances (1" min. and $1\frac{1}{2}$ to 2" for tall, irregular structures)

Figure E-6

161

Tolerances
for Interfacing

Definition. Interface tolerances are those required for joining of different materials and for accommodating the relative movements expected between such materials during the service life of the building.

Example. Tolerances for window and door openings, glazing gaskets, and reglets.

Selection. Where the matching of the manufactured materials is dependent on work executed at the construction site, interface tolerances should be equivalent to erection tolerances. Where the execution is independent of site work, tolerances should closely match the standard tolerances for the materials to be joined plus an appropriate allowance for differential volume changes between the materials being joined. Special tolerances for the planeness of concrete surfaces at the glass face should be provided when glazing is performed directly into the concrete. Close tolerances must govern the physical dimensions and alignment of the groove when neoprene gaskets are used in glazing.

Drawing symbols provide two definite advantages: the increased speed with which the draftsman is able to execute his work; and the positive description which they connote to the drawing user.

Drawing Symbols

The following section describes all of the commonly used drafting symbols. The various types of symbols common to architectural precast concrete are: (1) drawing, (2) graphic, (3) finish, (4) insert, (5) welding, and (6) mathematical. Any symbol which may be required and is not shown on these pages, must be described on the cover sheet. Many of the symbols can be quickly drawn with templates.

Symbols are employed to save time. If you prefer to describe any item or condition, symbols may be omitted.

Examples:

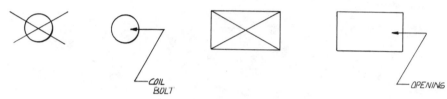

DRAWING SYMBOLS

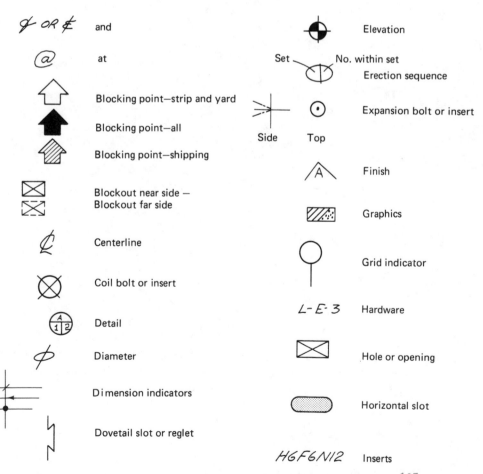

DRAWING SYMBOLS

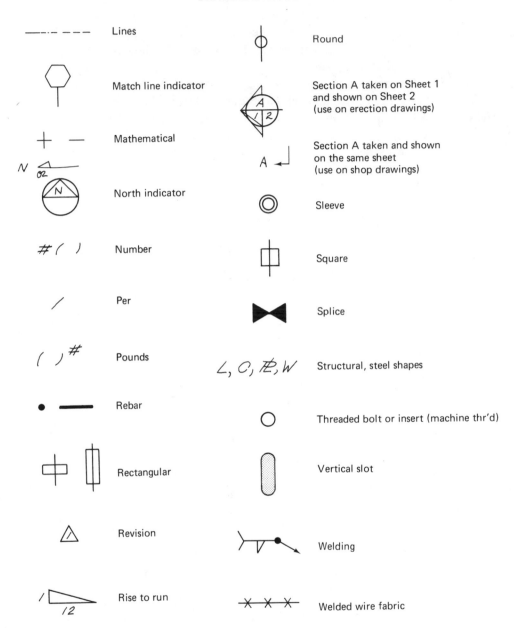

Graphic Symbols   Graphic symbols should be indicative of the material they represent. Clarify the symbol with a simple note if it is misleading or gives the impression of the wrong material.

Most symbols are different on elevations from those in section or on plans.

Material symbols or *hatching* need not be made with heavy line weights. In fact, their representation in rather large areas will appear more pleasing if the lineword "fades out" near the center, thus resembling a highlight on the surface, and it is given more emphasis near the edges. Make the so-called highlights non-geometric in shape, to eliminate any possibility of their appearing as construction features on the surface. The edges of hatched areas should be distinct. Contrast of adjacent materials is especially important on section views. To aid expression, hatch adjacent pieces on section views in different directions. Otherwise, use the same tone and technique for a material or member reappearing in different places throughout the section. If an overall gray pencil tone is used to indicate a predominant material, usually it is more tidy to apply the shading on the reverse side of the tracing.

If notes or dimensions must go into hatched areas, leave out enough of the linework so that the number or note can be inserted and can be read without difficulty.

GRAPHIC SYMBOLS				
Material	Elev.	Small Scale	Sect.	Small Scale
Block (all)				
Brick (all)				
CIP concrete	CIP			
Caulking				
Earth				
Fill				
Glass				
Gravel				
Grout				
Insulation				
Metal (not steel)				
Neoprene				
Precast concrete	1 2 3			
Rubber				
Sand				
Steel				
Stone (all)				
Wood (all)				

Any graphic symbols used other than those shown must be defined on the cover sheet of the specific project.

Standard Finish Symbols   The draftsman should give sufficent details or descriptions on his drawings to indicate clearly all exposed surfaces of the units and their respective finishes. This is particularly important for returns and interior finishes.

Finishes are indicated in the following manner. A triangle containing a letter designation shall be placed with its tip touching the surface to be indicated. Finishes shall be indicated where the surface appears as an edge.

If more than one finish occurs on a single surface, the triangle must be placed on the dimension line (as shown on the right side of the back view in the example below). All other views must then have a note referring to the view in which the limits of the two finishes are shown (see note on right side of top view). Finishes should be indicated on *all* panel edges of *all* views.

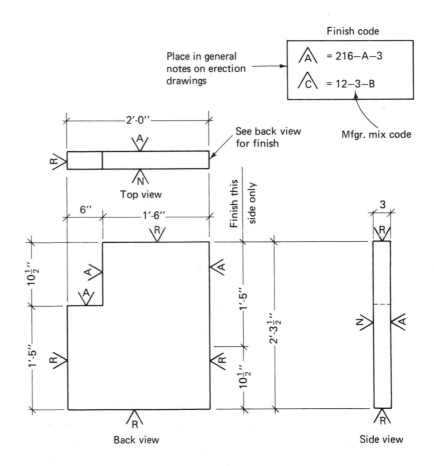

Standard Finish Symbol Nomenclature follows:

P/C FINISHES						
Type	Appearance	Graphic Notation	Method	When Done	Notes	Symbol
Acid etched	Fine textured exposed aggregate	——	Dip unit in or wash it with acid	After casting	Generally used for entire unit	/A\
Broom	Rough, lined surface	——	Stroke plastic concrete with a broom	During casting	Indicate direction of stroke	/B\
Exposed aggregate	Exposed stones raised above the matrix surface	Use in section only, typ.	One method is to apply retarder to formwork. Wash after stripping	Within 12 hours after stripping	Provide stop detail at point where E.A.C. finish meets smooth finish	/C\
Form liner	Repetitive design	⊓⊔⊓⊔	Affix liner to form face		Take care in detailing to align finish horizontally and vertically with adjacent panels	/D\
Fractured rib	Alternate strips of smooth and irregular concrete	⊓⊔⊓⊔	Affix ribbed form liner to mold. Break ribs from hardened concrete	2-3 weeks after stripping		/E\
Honed (ground)	Smooth exposed aggregate	——	Grind surface with # 24 – # 300 grits	Begin process $1\frac{1}{2}$ days after stripping		/F\
Polished (rubbed)	Smoother than honed	——	Hone first, apply matrix slurry, grind, buff with polishing compound	Begin process $1\frac{1}{2}$ days after stripping		/G\
Rough finish for grout bond	Rough, irregular	～～～	Varies	After casting		/H\

P/C FINISHES						
Type	Appearance	Graphic Notation	Method	When Done	Note	Symbol
Sandblasted	Shallow aggregate exposure leaving aggregates with a muted or frosted appearance		Bombard surface with sand particles driven by compressed air	2 days after stripping	For deep exposure, retarders or face-up water washing may be used initially	/J\
Screeded	Straight, flat, rough		Strike back surface off even with side rails	Immediately after casting		/K\
Smooth float	Smooth, not shiny or slick		Stroke plastic concrete with a mason's wood	Immediately after casting	Use when panel back is to receive insulation	/L\
Smooth form	Unmarked surface		Cast in carefully prepared form. Fill surface voids after stripping if required		Provide drips as required	/M\
Smooth steel trowel	Even, smooth		Work plastic concrete with a mason's steel trowel	After casting	Do not call for unless absolutely necessary	/N\
Tooled (bush hammer)	Rough exposed aggregate. Some fractured aggregates		Chip away surface cement paste with air driven chisel	2-3 weeks after stripping		/P\
Unfinished must be straight and flat	As cast, straight and flat		Varies		Use whenever possible	/R\
Stippled	Similar to a sand plaster wall		Roll surface with a latex paint roller following strike off and wood float finishing	Soon after casting	Normally used when interior face is to be painted	/S\

P/C FINISHES						
Type	Appearance	Graphic Notation	Method	When Done	Note	Symbol
As cast	As cast	————	Varies		Use whenever possible	△T
Veneer panel	Varies	————	Concrete cast with anchorage to veneer	At casting	Specify material (stone, brick, etc.)	△U

NOTE: Two concrete mixes with different colored matrixes exposed at the face of the panel should not be specified within one unit. No demarcation feature can prevent a white cement paste from leaking into a grey or vice versa. Such mixtures are acceptable (although at added cost) where one part of the panel (for instance, a spandrel part) is cast first of the one mix and, after curing, is cast into the total panel. This method has been successfully employed by proper sealing of the cold joint and by providing special protection to the initially cast concrete element.

INSERT SYMBOLS

Insert Symbols

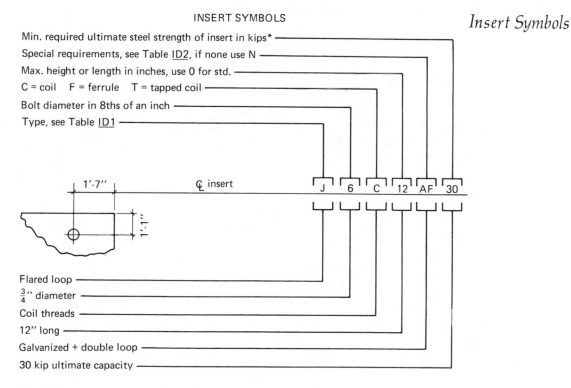

Min. required ultimate steel strength of insert in kips* ——————
Special requirements, see Table ID2, if none use N ——————
Max. height or length in inches, use 0 for std. ——————
C = coil   F = ferrule   T = tapped coil ——————
Bolt diameter in 8ths of an inch ——————
Type, see Table ID1 ——————

1'-7"   ℄ insert   J   6   C   12   AF   30

Flared loop ——————
¾" diameter ——————
Coil threads ——————
12" long ——————
Galvanized + double loop ——————
30 kip ultimate capacity ——————

*As limited by welding of wire anchors to coil or ferrule, diameter of wire anchors, and number of wire anchors attached. See Table ID3

Table ID1

TYPE	SYMBOL	INSERT	TYPE	SYMBOL	INSERT
A		Thin slab	J		Flared loop
B		Wing nut	K		Ferrule
C		Type "S"	L		
D		Loop	M		
E		Open coil	N		
F		Straight loop	P		
G		Coil rod anchor	R		
H		Coil tie	S		

Note: See various manufacturers' catalogs for complete information

Table ID2 SPECIAL REQUIREMENTS FOR INSERTS

A	Galvanized	G	Four struts
B	Stainless steel	H	Six struts
C	Cadium plated	J	
D	Non-corrosive for contact with form face	K	
		L	
E	Double loop	M	
F	Criss-cross double loop	N	None

Table ID3 APPROXIMATE CAPACITIES* OF ROUND WIRE USED WITH CONCRETE INSERTS

Wire Diameter Inch	Wire Grade	Approximate Yield Strength lbs.
0.218	C 1008	2000
0.223	C 1038	3900
0.225	C 1038	3700
0.240	C 1008	2900
0.260	C 1008	3550
0.281	C 1035	6000
0.306	C 1035	6900
0.340	C 1035	7500
0.375	C 1008	7450
0.440	C 1035	12000

*Table based upon typical data supplied by manufacturers of inserts
NOTE: Check manufacturer's catalog for wire sizes used with various insert types

+		Plus, positive (addition)	a^{-1}, a^{-2}	$\frac{1}{a}, \frac{1}{a^2}$	*Mathematical*
−		Minus, negative (subtraction)	$\sin^{-1}x$		*Symbols*
±	∓	Plus or minus, (minus or plus)	or	The angle whose sin is x	
×	·	Times, multiplied by (multiplication)	arc sin x		
÷	/	Divided by	Σ	Summation of	
( )	[ ] { }	Parentheses, brackets, braces	Δ	Increment of, difference	
=		Equals	‖	Absolute value of	
~	≈	Approximately equals	1.2345×10^6 =	1,234,500	
⇔		Corresponds to	1.2345×10^{-6} =	.0000012345	
≠		Does not equal	a! or ⌐a	Factorial a $(1 \cdot 2 \cdot 3 \cdots a)$	
>		Greater than	$\sqrt{x}$	Radical, square root of x	
<		Less than	$\sqrt[3]{x}, \sqrt[n]{x}$	Cube root of x, nth root of x	
≧		Greater than or equal to	i	$\sqrt{-1}$	
≦		Less than or equal to	∞	Infinity	
≡		Identical with	$\pi = 3.14159+$	Ratio of circumference to diameter of circle	
→	≐	Approaches as a limit	$e = 2.71828+$	Base of natural or napierian logarithms	
∝		Varies directly as	‖	Parallel to	
:		Is to (ratio)	⊥	Perpendicular to	
::		As, equals, so is	∠, ∠s	Angle, angles	
∴		Therefore	∟	Right angle	
a', a''		a-prime, a-double prime or a-second	△	Triangle	
a_1, a_2		a-sub one, a-sub two	○ ⊙	Circle	
a^2, a^3		a-squared, a-cubed	□	Square	
a^n		a raised to n^{th} power	°	Degree (angle, arc or temperature)	
		$\log_e a = 2.3026 \log_{10} a$	′	Feet or minutes	
		$\log_{10} a = 0.4343 \log_e a$	″	Inches or seconds	

Welding Symbols

Shop and erection drawings must provide specific instructions for the type, size, and length of welds, and their locations on the assembled piece. This information is usually given by means of welding symbols described in the American Welding Society booklet *Standard Welding Symbols*, AWS A2.0. The symbols in this system, commonly used, are shown in figure F-1.

Three basic parts are needed to form a welding symbol: an arrow pointing to the joint, a reference line upon which dimensional data is placed, and a basic weld symbol device indicating the weld type required.

A fourth part of the welding symbol, the tail, is used only when it is necessary to supply additional data such as specification, process, or detail references. An indication of specification references in the tail is necessary only when two or more electrode classes are required for the welding on a particular drawing. Since specification references usually determine the process, process references will be needed only for electrogas, electroslag, stud, or other kinds of welding where the electrode specification does not describe the process or method. When references are not needed to supplement the welding symbol, the tail is omitted.

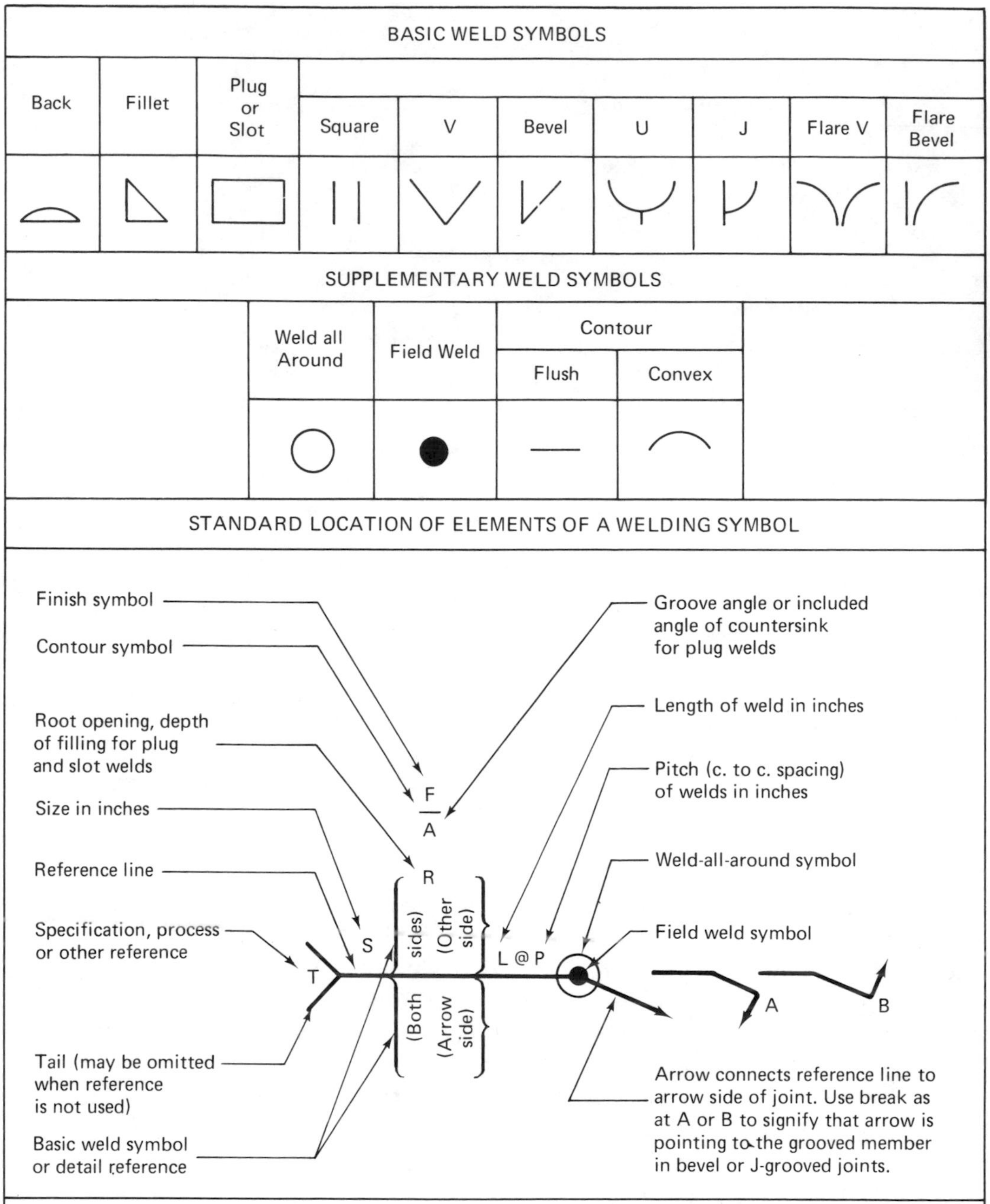

BASIC WELD SYMBOLS

Back	Fillet	Plug or Slot	Square	V	Bevel	U	J	Flare V	Flare Bevel
⌒	◺	▭	‖	V	⌵	⩁	⫫	⫝̸	⟋⎸

SUPPLEMENTARY WELD SYMBOLS

Weld all Around	Field Weld	Contour		
		Flush	Convex	
◯	●	—	⌒	

STANDARD LOCATION OF ELEMENTS OF A WELDING SYMBOL

Finish symbol

Contour symbol

Root opening, depth of filling for plug and slot welds

Size in inches

Reference line

Specification, process or other reference

Tail (may be omitted when reference is not used)

Basic weld symbol or detail reference

Groove angle or included angle of countersink for plug welds

Length of weld in inches

Pitch (c. to c. spacing) of welds in inches

Weld-all-around symbol

Field weld symbol

Arrow connects reference line to arrow side of joint. Use break as at A or B to signify that arrow is pointing to the grooved member in bevel or J-grooved joints.

F / A

R

S

(Both sides) (Other side)

(Both (Arrow side)

T

L @ P

A   B

Note:
 Size, weld symbol, length of weld and spacing must read in that order from left to right along the reference line. Neither orientation of reference line nor location of the arrow alter this rule.

 The perpendicular leg of ◺ , V , ⫫ , ⟋⎸ weld symbols must be at left.

 Arrow and Other Side welds are of the same size unless otherwise shown.

 Symbols apply between abrupt changes in direction of welding unless governed by the "all round" symbol or otherwise dimensioned.

 These symbols do not explicitly provide for the case that frequently occurs in structural work, where duplicate material (such as stiffeners) occurs on the far side of a web or gusset plate. The fabricating industry has adopted this convention; that when the billing of the detail material discloses the identity of far side with near side, the welding shown for the near side shall also be duplicated on the far side.

Figure F-1

The symbols for welds should be made large enough to be easily recognized and understood. Welding symbol templates may be purchased from suppliers of drafting equipment.

When welding reinforcement, weldability characteristics should be known. Depending upon the carbon and manganese content of the reinforcement, special welding procedures may be necessary. Unless otherwise required, it is recommended that only Grade 40 or Grade 60 reinforcing bars, with carbon contents not exceeding 0.50 percent and manganese contents not exceeding 1.30 percent, be welded. Only low hydrogen electrodes such as AWS Class E7015 or E7016 should be employed.

It is important not to weld reinforcing bars within eight bar diameters of a cold bend. This results in crystallization and unpredictable behavior of the reinforcing bar at the bend. (See figure F-2.) Likewise, tack welding may produce similar crystallization, and reduced bar strength, and should be carefully performed when required.

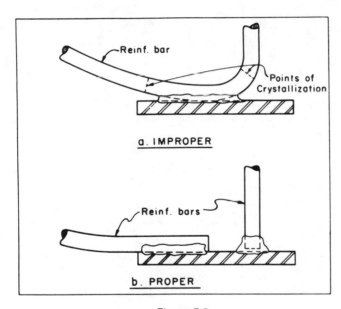

Figure F-2

Dead

        Unit weight
        permanently affixed material

Live

        Moveable
          Furniture
          Snow
          Water (pool)
        Moving
          People
          Vehicles
          Cranes
          Equipment

        Impact

         Material being dropped

        Lateral
          Wind (pressure and suction)
          Earthquake (seismic)
        Longitudinal

         Stopping craneway or vehicle

Dead and Live Load Distribution		
Type		Example
Concentrated or point (lbs)	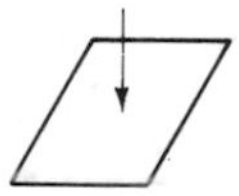	Column on footing
Line (PLF)	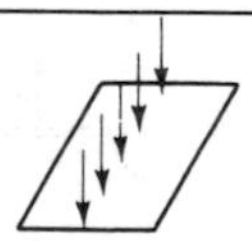	Bearing wall on a floor
Uniformly distributed (PSF)	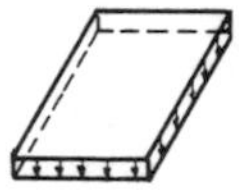	Topping, snow, ceiling

Superimposed load = total of all externally applied dead and live loads

When detailing a precast panel, it is important that the size of facing aggregate and its effect on reinforcing, hardware, and panel shape be given prime consideration. The exact aggregate size should be obtained from the supplier.

Gradation standards for aggregates for precast exposed aggregate products have varied widely. In the past, each producer of aggregates has more or less established his own sizing schedule or used standards of the terrazzo industry. This has resulted in confusion to both Precasters and Architects. In 1967, the *National Quartz Producers Council* was formed primarily to establish uniform standards for sizing hard aggregates used in exposed aggregate work. These size specifications are given in Table H-1:

Table H-1. NATIONAL QUARTZ PRODUCERS COUNCIL
SIZE SPECIFICATIONS

Size Designation	Percent Passing			
	$1\frac{3}{8}'' \times \frac{7}{8}''$ "D"	$\frac{7}{8}'' \times \frac{1}{2}''$ "C"	$\frac{1}{2}'' \times \frac{1}{4}''$ "B"	$\frac{1}{4}'' \times \frac{3}{32}''$ "A"
$1\frac{1}{2}$	100			
$1\frac{3}{8}$	95-100			
1	30-60	100		
$\frac{7}{8}$	20-40	95-100		
$\frac{5}{8}$	0-10	30-50	100	
$\frac{1}{2}$		10-25	95-100	
$\frac{3}{8}$		0-10	40-70	100
$\frac{1}{4}$			10-20	95-100
$\frac{1}{8}$			0-10	15-35
$\frac{3}{32}$				0-10

Generally, problems begin to occur when the facing aggregate size exceeds 3/4 inch. When the size of this aggregate is not taken into consideration, one or more of the problems illustrated in figure H-1 may occur.

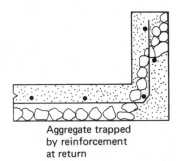

Aggregate trapped
by reinforcement
at return

Aggregate segregated
at insert or anchor

Aggregate trapped
by tightly spaced
reinforcement.
Center to center
spacing of reinforce-
ment must be considered

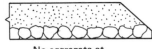

No aggregate at
quirk mitre. Quirk
dimension critical
(See page   )

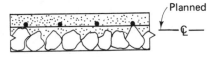

Hand placed reinforcement
mislocated

Figure H-1

PRECAST PANEL JOINTS

Type	Plan Example	Note
Butt		
False		Depth not to exceed width. Watch for reduced section!!
Inside mitre		
Quirk mitre		See For dimensions
Ship lap		Use when flame passage thru joint is a consideration. Problem may occur at intersection of horizontal and vertical joints and in erection sequence
Standard		Employ bevel if possible
Two stage (simple)		
Two stage (complex)		See "Architectural Precast Concrete Joint Details" report

Mitred corners are difficult to manufacture and erect within tolerances which are acceptable from either an appearance or jointing standpoint. Concrete cannot be cast to a 45° point because of the size of the aggregates. Therefore this edge must have a cut-off or quirk. The size of the quirk return should never be less than 3/4 inch, nor less than 1.5 times the maximum size of the aggregate used in the concrete mix.

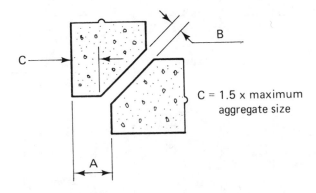

C = 1.5 x maximum aggregate size

Table J-1.

A (Quirk)	B (Joint)	C
1″	$\frac{3}{8}$″	$\frac{3}{4}$″
$1\frac{1}{4}$″	$\frac{3}{8}$″	1″
$1\frac{1}{2}$″	$\frac{3}{8}$″	$1\frac{1}{4}$″
$1\frac{3}{4}$″	$\frac{3}{8}$″	$1\frac{1}{2}$″
2″	$\frac{3}{8}$″	$1\frac{3}{4}$″
$1\frac{1}{4}$″	$\frac{1}{2}$″	$\frac{7}{8}$″
$1\frac{1}{2}$	$\frac{1}{2}$″	$1\frac{1}{8}$″
$1\frac{3}{4}$″	$\frac{1}{2}$″	$1\frac{3}{8}$″
2″	$\frac{1}{2}$″	$1\frac{5}{8}$″

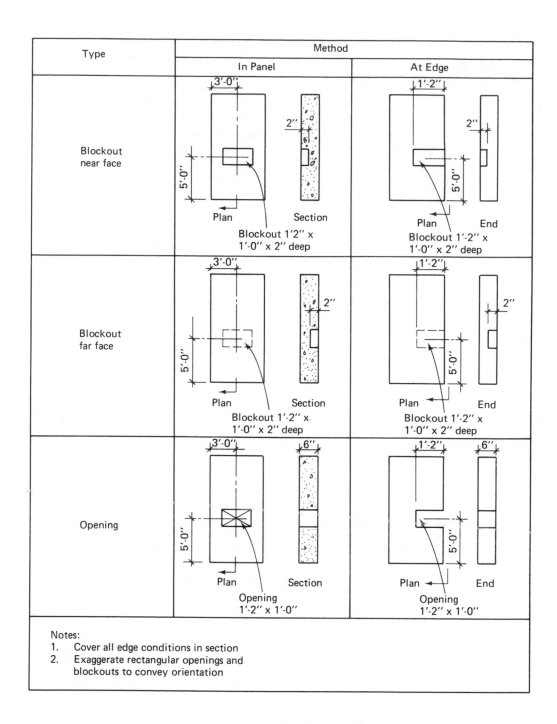

Type	Method	
	In Panel	At Edge
Blockout near face	3'-0" 2" 5'-0" Plan Section Blockout 1'2" x 1'-0" x 2" deep	1'-2" 2" 5'-0" Plan End Blockout 1'-2" x 1'-0" x 2" deep
Blockout far face	3'-0" 2" 5'-0" Plan Section Blockout 1'-2" x 1'-0" x 2" deep	1'-2" 2" 5'-0" Plan End Blockout 1'-2" x 1'-0" x 2" deep
Opening	3'-0" 6" 5'-0" Plan Section Opening 1'-2" x 1'-0"	1'-2" 6" 5'-0" Plan End Opening 1'-2" x 1'-0"

Notes:
1. Cover all edge conditions in section
2. Exaggerate rectangular openings and blockouts to convey orientation

RECOMMENDED MINIMUM BOLT PENETRATION

Coil	D	P
	$\frac{1}{2}''$	$1\frac{1}{2}''$
	$\frac{3}{4}''$	$2''$
	$1''$	$2\frac{1}{2}''$
	$1\frac{1}{4}''$	$2\frac{1}{2}''$
	$1\frac{1}{2}''$	$3''$
Machine	D	P
	$\frac{1}{2}''$	$1''$
	$\frac{5}{8}''$	$1\frac{1}{8}''$
	$\frac{3}{4}''$	$1\frac{1}{8}''$
	$1''$	$1\frac{1}{4}''$

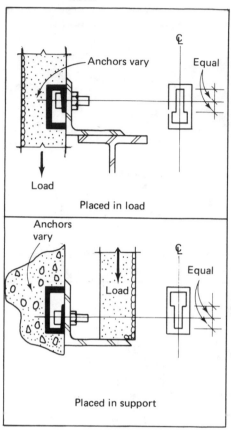

Adjustable (wedge type) Insert

Placed in load

Placed in support

NOTE: When using this type of insert, care should be taken to install with proper orientation. Use for lightly loaded connections only.

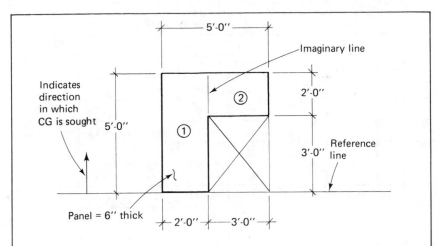

V = volume = length x width x depth
D = distance from chosen reference
    line to center of section

Section	V	D		V x D
1	5 cu ft	2.5′	=	12.5
2	3 cu ft	4.0′	=	12.0
	V total = 8 cu ft		VD total =	24.5

$$CG = \frac{VD_T}{V_T} = \frac{24.5}{8} = 3.062' \text{ from reference life}$$

To find horizontal CG repeat
operation using left side of
panel as new reference line

Table O FLAT PANEL SIZING AND CONNECTION LOCATION
TABLE [**NOT FOR DESIGN USE – DOES NOT CONSIDER
WARPAGE TOLERANCES** (See Appendix E, Table 2.)]

Thickness (in.)	Stripping and Flat Blocking at 2000 psi (ft)			Flat Erection* or Shipping 6000 psi (ft)			D** (ft)
	Ends	2 pt. pick	4 pt. pick	End	.207L	.292L	
2	6.5	16.5	33.0	9.0	12.5	15.5	9.0
$2\frac{1}{2}$	7.5	18.5	36.5	10.0	14.0	17.5	11.5
3	8.0	20.0	40.0	11.0	15.0	19.0	13.5
$3\frac{1}{2}$	9.0	21.5	43.5	12.0	16.5	21.0	16.0
4	9.5	23.0	46.5	13.0	17.5	22.0	18.5
$4\frac{1}{2}$	10.0	24.5	49.0	13.5	18.5	23.5	20.5
5	10.5	26.0	52.0	14.5	19.5	25.0	23.0
$5\frac{1}{2}$	11.0	27.0	54.5	15.0	20.5	26.0	25.0
6	11.5	28.5	57.0	16.0	21.5	27.0	27.5
$6\frac{1}{2}$	12.0	29.5	59.0	16.5	22.5	28.5	30.0
7	12.5	31.0	61.5	17.0	23.5	29.5	32.0
$7\frac{1}{2}$	13.0	32.0	63.5	17.5	24.0	30.5	34.5
8	13.5	33.0	65.5	18.5	25.0	31.5	37.0
$8\frac{1}{2}$	14.0	34.0	68.0	19.0	25.5	32.5	39.0
9	14.5	35.0	70.0	19.5	26.5	33.5	41.5
$9\frac{1}{2}$	15.0	36.0	71.5	20.0	27.0	34.5	43.5
10	15.5	37.0	73.5	20.5	28.0	35.0	46.0

*If length exceeds, ship vertical
**Using 25 psf wind pressure and satisfying
deflections of L/360

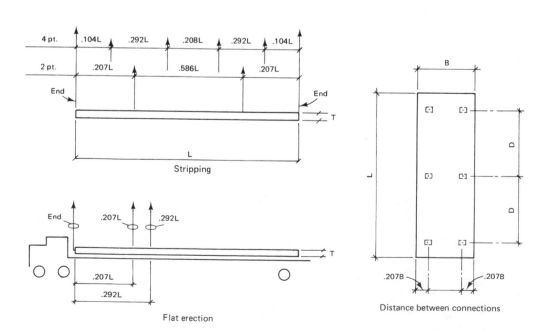

Stripping

Flat erection

Distance between connections

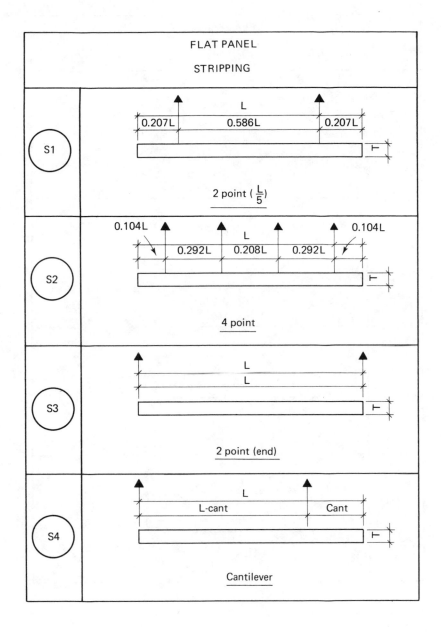

FLAT PANEL

STRIPPING

S1 — 2 point ($\frac{L}{5}$)

S2 — 4 point

S3 — 2 point (end)

S4 — Cantilever

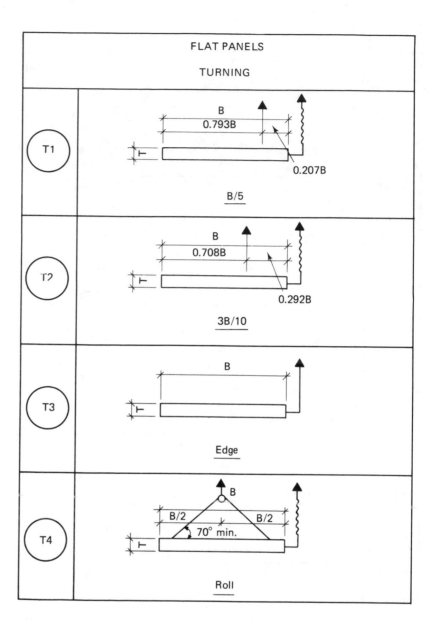

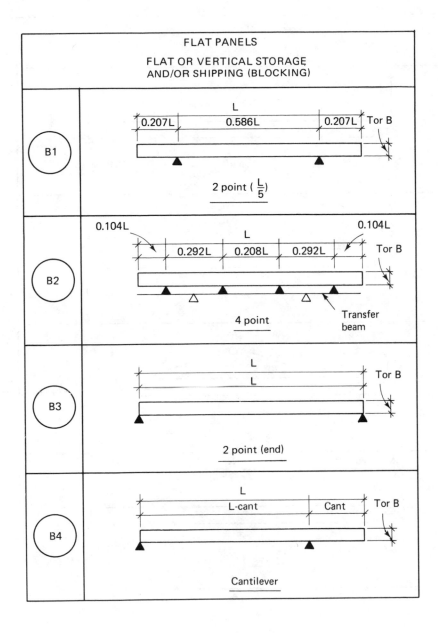

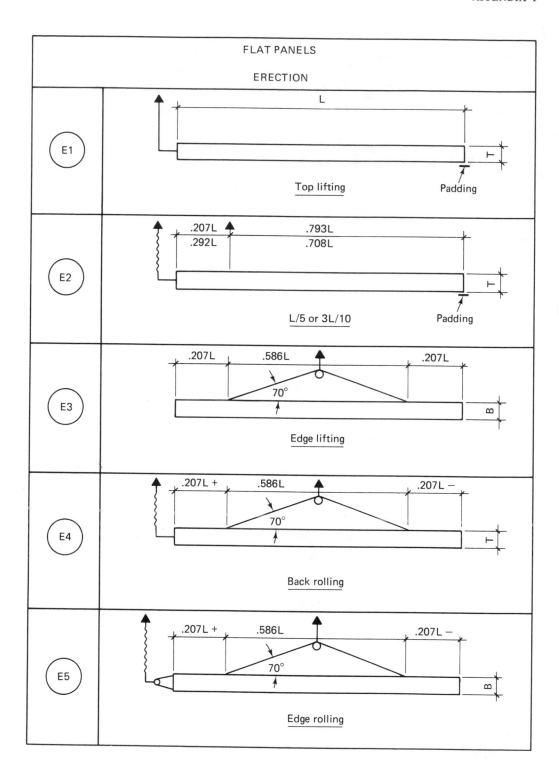

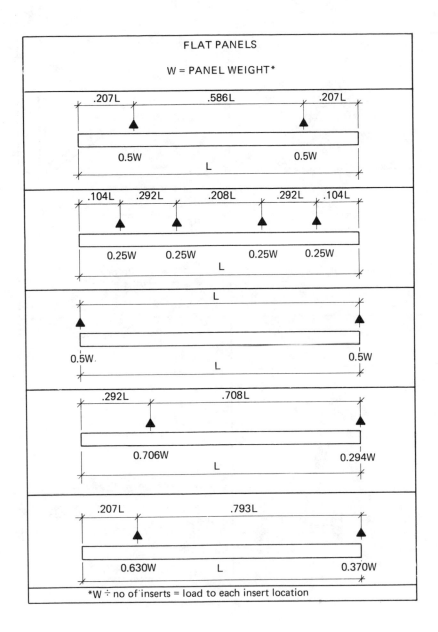

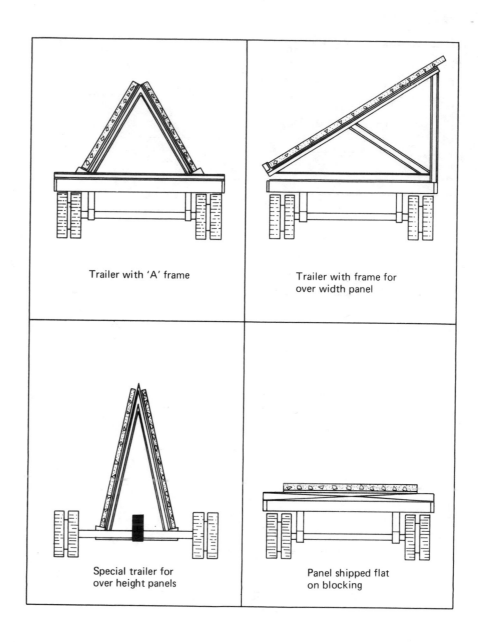

Trailer with 'A' frame

Trailer with frame for over width panel

Special trailer for over height panels

Panel shipped flat on blocking

A new method of designating wire sizes for concrete reinforcement was adopted in 1970 by the *American Society for Testing and Materials*. The new system applies to ASTM Designation A 82, *Standard Specifications for Cold-Drawn Steel Wire for Concrete Reinforcement*, and ASTM Designation A 185, *Standard Specification for Welded Steel Wire Fabric for Concrete Reinforcement*. In this system steel wire gage numbers are replaced by W-number size designations. Table S-1 shows a comparison of W-number wire sizes to nominal decimal diameters and to the steel wire gage system.

Transition to exclusive use of the new specifications will not be immediate and may require some period of time. Welded wire fabric may be specified during this transition period in accordance with ASTM A 185 by using either the former steel wire gages or by the new W-numbers.

Cross-sectional area is the basic element used in specifying wire size. Smooth wire sizes are specified by the letter *W* followed by a number indicating the cross-sectional area of the wire in hundredths of a square inch. For example: W16 denotes a smooth wire with cross-sectional area of 0.16 square inch: W5.5 identifies a smooth wire with cross-sectional area of 0.055 square inch, etc. Similarly, deformed wire sizes are specified by the letter *D* followed by a number which indicates hundredths of a square inch. For example: D10 is a deformed wire with a cross-sectional area of 0.10 square inch.

Welded wire fabric is usually denoted on design drawings as follows: WWF followed by spacings of longitudinal wires, then by transverse wires, and last by the sizes of longitudinal and transverse wires.

Spacings and sizes of wires in welded wire fabric are identified by *style*. A typical style designation is: 6 X 12—W16 X W8 (as shown in figure S-1).

This denotes a welded wire fabric in which:

    Spacing of longitudinal wires = 6 inches
    Spacing of transverse wires = 12 inches
    Size of longitudinal wires = W16 (0.16 square inch)
    Size of transverse wires = W8 (0.08 square inch)

A welded deformed wire fabric style would be noted in the same manner by substituting D-number wire sizes for the W-number wire sizes shown.

It is very important to note that the terms longitudinal and transverse are related to the manufacturing process and have *no* reference to the position of the wires in a concrete structure.

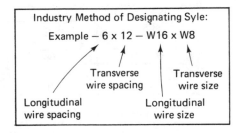

Figure S-1

COMPARISON BETWEEN NEW W-NUMBER WIRE SIZES
AND STEEL WIRE GAGES

W-NUMBER WIRE SIZES			STEEL WIRE GAGES		
Size Number	Area, Sq. in.	Nom. Dia., In.	Nom. Dia., In.	Area, Sq. In.	Gage Number
			0.0625	0.003	16
			0.072	0.004	15
W0.5	0.005	0.080	0.080	0.005	14
W0.7	0.007	0.0915	0.0915	0.007	13
W0.9	0.009	0.1055	0.1055	0.009	12
W1	0.01	0.113			
W1.1	0.011	0.1205	0.1205	0.011	11
W1.4	0.014	0.135	0.135	0.014	10
W1.5	0.015	0.138			
W1.7	0.017	0.148	0.148	0.017	9
W2	0.02	0.159			
W2.1	0.021	0.162	0.162	0.021	8
W2.5	0.025	0.178	0.177	0.025	7
W2.9	0.029	0.192	0.192	0.029	6
W3	0.03	0.195			
W3.4	0.034	0.207	0.207	0.034	5
W3.5	0.035	0.211			
W4	0.04	0.225	0.225	0.04	4
			0.235	0.043	3½
W4.5	0.045	0.240			
			0.244	0.047	3
W5	0.05	0.252	0.253	0.05	2½
			0.2625	0.054	2
W5.5	0.055	0.264			
			0.273	0.059	1½
W6	0.06	0.276			
			0.283	0.063	1
W6.5	0.065	0.288			
			0.295	0.068	1/0½
W7	0.07	0.298			
			0.3065	0.074	1/0
W7.5	0.075	0.309			
W8	0.08	0.319	0.319	0.08	2 0½
W8.5	0.085	0.329			
			0.331	0.086	2/0
W9	0.09	0.338			
W9.5	0.095	0.348	0.347	0.095	3/0½
W10	0.10	0.356			
			0.3625	0.103	3/0
W10.5	0.105	0.366			
W11	0.11	0.374			
			0.378	0.112	4/0½
W12	0.12	0.390			
			0.394	0.122	4/0
W14	0.14	0.422			
			0.4305	0.146	5/0
W16	0.16	0.451			
			0.4615	0.167	6/0
W18	0.18	0.478			
			0.490	0.189	7/0
W20	0.20	0.504			

Table S-1   *(Courtesy of Wire Reinforcement Institute)*

Table S-2. MATERIAL PROPERTIES WELDED WIRE FABRIC
Properties of common styles of welded wire fabric

	Style Designation	Spacing of Wires, in.		Size of Wires, AS & W gage		Sectional Area, sq.in. per ft		Weight, lb per 100 sq ft
		Longit.	Trans.	Longit.	Trans.	Longit.	Trans.	
Two-Way Types	2 x 2-10/10*	2	2	10	10	.086	.086	60
	2 x 2-12/12*	2	2	12	12	.052	.052	37
	2 x 2-14/14*	2	2	14	14	.030	.030	21
	3 x 3-8/8	3	3	8	8	.082	.082	58
	3 x 3-10/10	3	3	10	10	.057	.057	41
	3 x 3-12/12*	3	3	12	12	.035	.035	25
	3 x 3-14/14*	3	3	14	14	.020	.020	14
	4 x 4-3/3**	4	4	3	3	.140	.140	100
	4 x 4-4/4**	4	4	4	4	.120	.120	85
	4 x 4-6/6**	4	4	6	6	.087	.087	62
	4 x 4-8/8**	4	4	8	8	.062	.062	44
	4 x 4-10/10**	4	4	10	10	.043	.043	31
	4 x 4-12/12*	4	4	12	12	.026	.026	19
	6 x 6-0/0	6	6	0	0	.148	.148	107
	6 x 6-2/2	6	6	2	2	.108	.108	78
	6 x 6-4/4	6	6	4	4	.080	.080	58
	6 x 6-4/6	6	6	4	6	.080	.058	50
	6 x 6-6/6	6	6	6	6	.058	.058	42
	6 x 6-8/8	6	6	8	8	.041	.041	30
	6 x-10/10	6	6	10	10	.029	.029	21
One-Way Types	2 x 12-0/4	2	12	0	4	.443	.040	169
	2 x 12-2/6	2	12	2	6	.325	.029	124
	2 x 12-4/8	2	12	4	8	.239	.021	91
	2 x 12-6/10	2	12	6	10	.174	.014	66
	2 x 12-8/12	2	12	8	12	.124	.009	46
	3 x 12-0/4	3	12	0	4	.295	.040	119
	3 x 12-2/6	3	12	2	6	.216	.029	87
	3 x 12-4/8	3	12	4	8	.159	.021	64
	3 x 12-6/10	3	12	6	10	.116	.014	46
	3 x 12-8/12	3	12	8	12	.082	.009	32
	4 x 8-8/12	4	8	8	12	.062	.013	27
	4 x 8-10/12	4	8	10	12	.043	.013	20
	4 x 12-0/4	4	12	0	4	.221	.040	94
	4 x 12-2/6	4	12	2	6	.162	.029	69
	4 x 12-4/8	4	12	4	8	.120	.021	51
	4 x 12-6/10	4	12	6	10	.087	.014	36
	4 x 12-10/12	4	12	10	12	.043	.009	19
	6 x 12-00/4	6	12	00	4	.172	.040	78
	6 x 12-0/4	6	12	0	4	.148	.040	69
	6 x 12-2/2	6	12	2	2	.108	.054	59
	6 x 12-4/4	6	12	4	4	.080	.040	44
	6 x 12-6/6	6	12	6	6	.058	.029	32

*Usually furnished only in galvanized wire.
**Most common sizes for architectural precast concrete.

SECTIONAL AREA AND WEIGHT OF WELDED WIRE FABRIC

(Area—sq. in. per ft. of width for various spacings)

Wire Size Number Smooth	Deformed	Nominal Diameter, Inches	Nominal Weight Lbs./Lin. Ft.	2″	3″	4″	6″	8″	10″	12″
W31	D31	0.628	1.054	1.86	1.24	.93	.62	.465	.372	.31
W30	D30	0.618	1.020	1.80	1.20	.90	.60	.45	.36	.30
W28	D28	0.597	.952	1.68	1.12	.84	.56	.42	.336	.28
W26	D26	0.575	.934	1.56	1.04	.78	.52	.39	.312	.26
W24	D24	0.553	.816	1.44	.96	.72	.48	.36	.288	.24
W22	D22	0.529	.748	1.32	.88	.66	.44	.33	.264	.22
W20	D20	0.504	.680	1.20	.80	.60	.40	.30	.24	.20
W18	D18	0.478	.612	1.08	.72	.54	.36	.27	.216	.18
W16	D16	0.451	.544	.96	.64	.48	.32	.24	.192	.16
W14	D14	0.422	.476	.84	.56	.42	.28	.21	.168	.14
W12	D12	0.390	.408	.72	.48	.36	.24	.18	.144	.12
W11	D11	0.374	.374	.66	.44	.33	.22	.165	.132	.11
W10.5		0.366	.357	.63	.42	.315	.21	.157	.126	.105
W10	D10	0.356	.340	.60	.40	.30	.20	.15	.12	.10
W9.5		0.348	.323	.57	.38	.285	.19	.142	.114	.095
W9	D9	0.338	.306	.54	.36	.27	.18	.135	.108	.09
W8.5		0.329	.289	.51	.34	.255	.17	.127	.102	.085
W8	D8	0.319	.272	.48	.32	.24	.16	.12	.096	.08
W7.5		0.309	.255	.45	.30	.225	.15	.112	.09	.075
W7	D7	0.298	.238	.42	.28	.21	.14	.105	.084	.07
W6.5		0.288	.221	.39	.26	.195	.13	.097	.078	.065
W6	D6	0.276	.204	.36	.24	.18	.12	.09	.072	.06
W5.5		0.264	.187	.33	.22	.165	.11	.082	.066	.055
W5	D5	0.252	.170	.30	.20	.15	.10	.075	.06	.05
W4.5		0.240	.153	.27	.18	.135	.09	.067	.054	.045
W4	D4	0.225	.136	.24	.16	.12	.08	.06	.048	.04
W3.5		0.211	.119	.21	.14	.105	.07	.052	.042	.035
W3		0.195	.102	.18	.12	.09	.06	.045	.036	.03
W2.9		0.192	.098	.174	.116	.087	.058	.043	.035	.029
W2.5		0.178	.085	.15	.10	.075	.05	.037	.03	.025
W2.1		0.162	.070	.126	.084	.063	.042	.031	.025	.021
W2		0.159	.068	.12	.08	.06	.04	.03	.024	.02
W1.5		0.138	.051	.09	.06	.045	.03	.022	.018	.015
W1.4		0.135	.049	.084	.056	.042	.028	.021	.017	.014

Table S-3   *(Courtesy of Wire Reinforcement Institute)*

> **NOTE:** Wire sizes other than those listed above may be produced provided the quantity required is sufficient to justify manufacture.

SECTIONAL AREAS OF STEEL REINFORCING

Wire Size Number (W or D Numbers) Center to Center Spacing — in.					Steel Area Square Inches Per Foot	Center to Center Spacing — in. Rebar Size Number		
2"	3"	4"	6"	12"		#3	#4	#5
				2	0.02			
			1.5	3	0.03			
			2	4	0.04			
			2.5	5	0.05			
	1.5	2	3	6	0.06			
			3.5	7	0.07	18		
	2		4	8	0.08	16½		
1.5		3	4.5	9	0.09	14½		
	2.5		5	10	0.10	13		
			5.5	11	0.11	12		
2	3	4	6	12	0.12	11		
			6.5		0.13	10	18	
	3.5		7	14	0.14	9½	17	
2.5		5	7.5		0.15	9	16	
	4		8	16	0.16	8½	15	
			8.5		0.17	8	14	
3	4.5	6	9	18	0.18	7½	13	
			9.5		0.19	7	12½	
	5		10	20	0.20	6½	12	18
3.5		7	10.5		0.21		11½	17½
	5.5		11	22	0.22	6	11	16½
					0.23		10½	16
4	6	8	12	24	0.24	5½	10	15½
					0.25		9½	15
	6.5			26	0.26	5		14
4.5		9			0.27		9	13½
	7		14	28	0.28	4½	8½	13
					0.29			
5	7.5	10		30	0.30		8	12½
				31	0.31			12
	8		16		0.32	4	7½	11½
5.5		11			0.33			
	8.5				0.34		7	11
					0.35			
6	9	12	18		0.36			10½
					0.37	3½	6½	10
	9.5				0.38			
6.5					0.39			9½
	10		20		0.40		6	
					0.41			9
7	10.5	14			0.42			
					0.43			
	11		22		0.44	3	5½	8½
7.5					0.45			
					0.46			8
					0.47			
8	12	16	24		0.48		5	7½
					0.49			
					0.50			
8.5					0.51			
			26		0.52			
					0.53			
9		18			0.54		4½	7
					0.55			
	14		28		0.56			6½
9.5					0.57			
					0.58			
					0.59			
10		20	30		0.60		4	

Table S-4   *(Courtesy of Wire Reinforcement Institute)*

> **NOTE:** The above table is based on equal allowable steel stresses. If comparing different grades of steel, the designer should compensate accordingly. For instance, f_y or Grade 40 or 50 steel is 40,000 psi whereas welded wire fabric and Grade 60 steel have a value of 60,000 psi for f_y. The weight saving of welded wire fabric, for instance, may be significant and should be considered when comparing with Grades 40 and 50 steel.

Table T-1. ASTM STANDARD REINFORCING BARS				
		Nominal Dimensions — Round Sections		
Bar Size Designation	Weight Pounds per foot	Diameter* Inches	Cross-sectional Area sq. inches	Perimeter inches
# 3	.376	.375	.11	1.178
# 4	.668	.500	.20	1.571
# 5	1.043	.625	.31	1.963
# 6	1.502	.750	.44	2.356
# 7	2.044	.875	.60	2.749
# 8	2.670	1.000	.79	3.142
# 9	3.400	1.128	1.00	3.544
#10	4.303	1.270	1.27	3.990
#11	5.313	1.410	1.56	4.430
#14	7.65	1.693	2.25	5.32
#18	13.60	2.257	4.00	7.09

*Not to be overlooked in dimensioning is that reinforcing bars have deformations that add $\frac{1}{8}$ inch or more to the nominal diameter of a reinforcing bar as shown below:

Reinforcing bar deformations

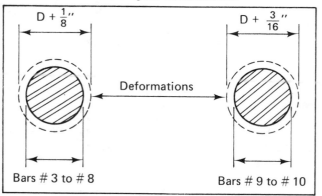

MATERIAL PROPERTIES
REINFORCING BARS
Areas and perimeters of reinforcing bar combinations

Areas, A_s (or A'_s) (top) sq. in.
Perimeters, Σo (bottom) in.
Columns headed [0] [5] contain data for bars of one size in groups of one to ten.
Columns headed [1][2][3][4][5] contain data for bars of two sizes with from one to five of each size.
For bars of one size: Σo = sum of perimeters
For bars of two sizes: $\Sigma o = \dfrac{4A}{D}$
where D = largest bar dia.

Each cell shows Area (top) / Perimeter (bottom).

Size	n	Size'	0	5	Size''	1	2	3	4	5	Size'''	1	2	3	4	5
#4	1	#4	0.20 / 1.6	1.20 / 9.4	#3	0.31 / 2.5	0.42 / 3.4	0.53 / 4.2	0.64 / 5.1	0.75 / 6.0						
	2		0.40 / 3.1	1.40 / 11.0		0.51 / 4.1	0.62 / 5.0	0.73 / 5.8	0.84 / 6.7	0.95 / 7.6						
	3		0.60 / 4.7	1.60 / 12.6		0.71 / 5.7	0.82 / 6.6	0.93 / 7.4	1.04 / 8.3	1.15 / 9.2						
	4		0.80 / 6.3	1.80 / 14.1		0.91 / 7.3	1.02 / 8.2	1.13 / 9.0	1.24 / 9.9	1.35 / 10.8						
	5		1.00 / 7.9	2.00 / 15.7		1.11 / 8.9	1.22 / 9.8	1.33 / 10.6	1.44 / 11.5	1.55 / 12.4						
#5	1	#5	0.31 / 2.0	1.86 / 11.8	#4	0.51 / 3.3	0.71 / 4.5	0.91 / 5.8	1.11 / 7.1	1.31 / 8.4	#3	0.42 / 2.7	0.53 / 3.4	0.64 / 4.1	0.75 / 4.8	0.86 / 5.5
	2		0.62 / 3.9	2.17 / 13.7		0.82 / 5.2	1.02 / 6.5	1.22 / 7.8	1.42 / 9.1	1.62 / 10.4		0.73 / 4.7	0.84 / 5.4	0.95 / 6.1	1.06 / 6.8	1.17 / 7.5
	3		0.93 / 5.9	2.48 / 15.7		1.13 / 7.2	1.33 / 8.5	1.53 / 9.8	1.73 / 11.1	1.93 / 12.4		1.04 / 6.7	1.15 / 7.4	1.26 / 8.1	1.37 / 8.8	1.48 / 9.5
	4		1.24 / 7.9	2.79 / 17.7		1.44 / 9.2	1.64 / 10.5	1.84 / 11.8	2.04 / 13.1	2.24 / 14.3		1.35 / 8.6	1.46 / 9.3	1.57 / 10.0	1.68 / 10.8	1.79 / 11.5
	5		1.55 / 9.8	3.10 / 19.6		1.75 / 11.2	1.95 / 12.5	2.15 / 13.8	2.35 / 15.0	2.55 / 16.3		1.66 / 10.6	1.77 / 11.3	1.88 / 12.0	1.99 / 12.7	2.10 / 13.4
#6	1	#6	0.44 / 2.4	2.64 / 14.1	#5	0.75 / 4.0	1.06 / 5.7	1.37 / 7.3	1.68 / 9.0	1.99 / 10.6	#4	0.64 / 3.4	0.84 / 4.5	1.04 / 5.5	1.24 / 6.6	1.44 / 7.7
	2		0.88 / 4.7	3.08 / 16.5		1.19 / 6.3	1.50 / 8.0	1.81 / 9.7	2.12 / 11.3	2.43 / 13.0		1.08 / 5.8	1.28 / 6.8	1.48 / 7.9	1.68 / 9.0	1.88 / 10.0
	3		1.32 / 7.1	3.52 / 18.8		1.63 / 8.7	1.94 / 10.3	2.25 / 12.0	2.56 / 13.7	2.87 / 15.3		1.52 / 8.1	1.72 / 9.2	1.92 / 10.2	2.12 / 11.3	2.32 / 12.4
	4		1.76 / 9.4	3.96 / 21.2		2.07 / 11.0	2.38 / 12.7	2.69 / 14.3	3.00 / 16.0	3.31 / 17.7		1.96 / 10.5	2.16 / 11.5	2.36 / 12.6	2.56 / 13.7	2.76 / 14.7
	5		2.20 / 11.8	4.40 / 23.6		2.51 / 13.4	2.82 / 15.0	3.13 / 16.7	3.44 / 18.3	3.75 / 20.0		2.40 / 12.8	2.60 / 13.9	2.80 / 14.9	3.00 / 16.0	3.20 / 17.1
#7	1	#7	0.60 / 2.7	3.60 / 16.5	#6	1.04 / 4.8	1.48 / 6.8	1.92 / 8.8	2.36 / 10.8	2.80 / 12.8	#5	0.91 / 4.2	1.22 / 5.6	1.53 / 7.0	1.84 / 8.4	2.15 / 9.8
	2		1.20 / 5.5	4.20 / 19.2		1.64 / 7.5	2.08 / 9.5	2.52 / 11.5	2.96 / 13.5	3.40 / 15.5		1.51 / 6.9	1.82 / 8.3	2.13 / 9.7	2.44 / 11.2	2.75 / 12.6
	3		1.80 / 8.2	4.80 / 22.0		2.24 / 10.2	2.68 / 12.3	3.12 / 14.3	3.56 / 16.3	4.00 / 18.3		2.11 / 9.6	2.42 / 11.1	2.73 / 12.5	3.04 / 13.9	3.35 / 15.3
	4		2.40 / 11.0	5.40 / 24.7		2.84 / 13.0	3.28 / 15.0	3.72 / 17.0	4.16 / 19.0	4.60 / 21.0		2.71 / 12.4	3.02 / 13.8	3.33 / 15.2	3.64 / 16.6	3.95 / 18.1
	5		3.00 / 13.7	6.00 / 27.5		3.44 / 15.7	3.88 / 17.7	4.32 / 19.7	4.76 / 21.8	5.20 / 23.8		3.31 / 15.1	3.62 / 16.5	3.93 / 18.0	4.24 / 19.4	4.55 / 20.8
#8	1	#8	0.79 / 3.1	4.74 / 18.9	#7	1.39 / 5.6	1.99 / 8.0	2.59 / 10.4	3.19 / 12.8	3.79 / 15.2	#6	1.23 / 4.9	1.67 / 6.7	2.11 / 8.4	2.55 / 10.2	2.99 / 12.0
	2		1.58 / 6.3	5.53 / 22.0		2.18 / 8.7	2.78 / 11.1	3.38 / 13.5	3.98 / 15.9	4.58 / 18.3		2.02 / 8.1	2.46 / 9.8	2.90 / 11.6	3.34 / 13.4	3.78 / 15.1
	3		2.37 / 9.4	6.32 / 25.1		2.97 / 11.9	3.57 / 14.3	4.17 / 16.7	4.77 / 19.1	5.37 / 21.5		2.81 / 11.2	3.25 / 13.0	3.69 / 14.8	4.13 / 16.5	4.57 / 18.3
	4		3.16 / 12.6	7.11 / 28.3		3.76 / 15.0	4.36 / 17.4	4.96 / 19.8	5.56 / 22.2	6.16 / 24.6		3.60 / 14.4	4.04 / 16.2	4.48 / 17.9	4.92 / 19.7	5.36 / 21.4
	5		3.95 / 15.7	7.90 / 31.4		4.55 / 18.2	5.15 / 20.6	5.75 / 23.0	6.35 / 25.4	6.95 / 27.8		4.39 / 17.6	4.83 / 19.3	5.27 / 21.1	5.71 / 22.8	6.15 / 24.6
#9	1	#9	1.00 / 3.5	6.00 / 21.3	#8	1.79 / 6.3	2.58 / 9.1	3.37 / 12.0	4.16 / 14.8	4.95 / 17.6	#7	1.60 / 5.7	2.20 / 7.8	2.80 / 9.9	3.40 / 12.1	4.00 / 14.2
	2		2.00 / 7.1	7.00 / 24.8		2.79 / 9.9	3.58 / 12.7	4.37 / 15.5	5.16 / 18.3	5.95 / 21.1		2.60 / 9.2	3.20 / 11.3	3.80 / 13.5	4.40 / 15.6	5.00 / 17.7
	3		3.00 / 10.6	8.00 / 28.4		3.79 / 13.4	4.58 / 16.2	5.37 / 19.0	6.16 / 21.8	6.95 / 24.6		3.60 / 12.8	4.20 / 14.9	4.80 / 17.0	5.40 / 19.1	6.00 / 21.3
	4		4.00 / 14.2	9.00 / 31.9		4.79 / 17.0	5.58 / 19.8	6.37 / 22.6	7.16 / 25.4	7.95 / 28.2		4.60 / 16.3	5.20 / 18.4	5.80 / 20.6	6.40 / 22.7	7.00 / 24.8
	5		5.00 / 17.7	10.00 / 35.5		5.79 / 20.5	6.58 / 23.3	7.37 / 26.1	8.16 / 28.9	8.95 / 31.7		5.60 / 19.9	6.20 / 22.0	6.80 / 24.1	7.40 / 26.2	8.00 / 28.4
#10	1	#10	1.27 / 4.0	7.62 / 24.0	#9	2.27 / 7.1	3.27 / 10.3	4.27 / 13.4	5.27 / 16.6	6.27 / 19.7	#8	2.06 / 6.5	2.85 / 9.0	3.64 / 11.5	4.43 / 14.0	5.22 / 16.4
	2		2.54 / 8.0	8.89 / 27.9		3.54 / 11.2	4.54 / 14.3	5.54 / 17.4	6.54 / 20.6	7.54 / 23.7		3.33 / 10.5	4.12 / 13.0	4.91 / 15.5	5.70 / 18.0	6.49 / 20.4
	3		3.81 / 12.0	10.16 / 31.9		4.81 / 15.2	5.81 / 18.3	6.81 / 21.4	7.81 / 24.6	8.81 / 27.7		4.60 / 14.5	5.39 / 17.0	6.18 / 19.5	6.97 / 22.0	7.76 / 24.4
	4		5.08 / 16.0	11.43 / 35.9		6.08 / 19.2	7.08 / 22.3	8.08 / 25.4	9.08 / 28.6	10.08 / 31.7		5.87 / 18.5	6.66 / 21.0	7.45 / 23.5	8.24 / 26.0	9.03 / 28.4
	5		6.35 / 20.0	12.70 / 39.9		7.35 / 23.2	8.35 / 26.3	9.35 / 29.4	10.35 / 32.6	11.35 / 35.7		7.14 / 22.5	7.93 / 25.0	8.72 / 27.5	9.51 / 30.0	10.30 / 32.4
#11	1	#11	1.56 / 4.4	9.36 / 26.6	#10	2.83 / 8.0	4.10 / 11.6	5.37 / 15.2	6.64 / 18.8	7.91 / 22.4	#9	2.56 / 7.3	3.56 / 10.1	4.56 / 12.9	5.56 / 15.8	6.56 / 18.6
	2		3.12 / 8.9	10.92 / 31.0		4.39 / 12.5	5.66 / 16.1	6.93 / 19.7	8.20 / 23.3	9.47 / 26.9		4.12 / 11.7	5.12 / 14.5	6.12 / 17.4	7.12 / 20.2	8.12 / 23.0
	3		4.68 / 13.3	12.48 / 35.4		5.95 / 16.9	7.22 / 20.5	8.49 / 24.1	9.76 / 27.7	11.03 / 31.3		5.68 / 16.1	6.68 / 19.0	7.68 / 21.8	8.68 / 24.6	9.68 / 27.5
	4		6.24 / 17.7	14.04 / 39.9		7.51 / 21.3	8.78 / 24.9	10.05 / 28.5	11.32 / 32.1	12.59 / 35.7		7.24 / 20.5	8.24 / 23.4	9.24 / 26.2	10.24 / 29.1	11.24 / 31.9
	5		7.80 / 22.2	15.60 / 44.3		9.07 / 25.7	10.34 / 29.3	11.61 / 32.9	12.88 / 36.5	14.15 / 40.1		8.80 / 25.0	9.80 / 27.8	10.80 / 30.6	11.80 / 33.5	12.80 / 36.3

Table T-2   *(Courtesy of Concrete Reinforcing Steel Institute)*

ACI STANDARD HOOKS

All specific sizes recommended by CRSI below meet requirements of ACI 318-71

RECOMMENDED END HOOKS
All Grades

D = 6d for #3 through #8
D = 8d for #9, #10, and #11
D = 10d for #14 and #18

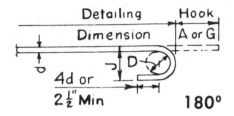

180°

Bar Size	180° HOOKS*		90° HOOKS
	A or G	J	A or G
#3	5	3	6
#4	6	4	8
#5	7	5	10
#6	8	6	1-0
#7	10	7	1-2
#8	11	8	1-4
#9	1-3	11¼	1-7
#10	1-5	1-0¾	1-10
#11	1-7	1-2¼	2-0
#14	2-2	1-8½	2-7
#18	2-11	2-3	3-5

*With Grade 40 only, where available depth is limited, bars may be bent with D = 5d for #3 through #11.

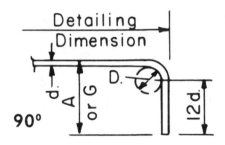

90°

STIRRUP AND TIE HOOKS

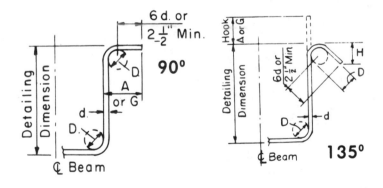

STIRRUPS
(TIES SIMILAR)

STIRRUP AND TIE HOOK DIMENSIONS
Grades 40-50-60 ksi

Bar Size	D (in.)	90° Hook	135° Hook	
		Hook A or G	Hook A or G	H Approx.
#3	1½	4	4	2½
#4	2	4½	4½	3
#5	2½	6	5½	3¾

LOCATION OF MAIN REINFORCEMENT

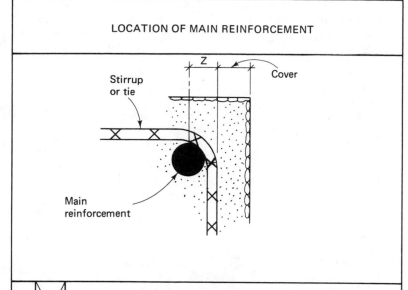

		Z Dimension		
		Stirrup or Tie Size		
		# 3	# 4	# 5
Main Reinforcement Size	# 4	$\frac{7}{8}''$	$1\frac{1}{16}''$	$1\frac{1}{4}''$
	# 5	$\frac{7}{8}''$	$1\frac{1}{8}''$	$1\frac{5}{16}''$
	# 6	$\frac{15}{16}''$	$1\frac{3}{16}''$	$1\frac{3}{8}''$
	# 7	$1''$	$1\frac{3}{16}''$	$1\frac{7}{16}''$
	# 8	$1\frac{1}{16}''$	$1\frac{1}{4}''$	$1\frac{7}{16}''$
	# 9	$1\frac{1}{16}''$	$1\frac{5}{16}''$	$1\frac{1}{2}''$
	# 10	$1\frac{1}{8}'$	$1\frac{5}{16}''$	$1\frac{1}{2}''$
	# 11	$1\frac{3}{16}''$	$1\frac{3}{8}''$	$1\frac{9}{16}''$

HOT-ROLLED STRUCTURAL STEEL SHAPE DESIGNATIONS

New Designation	Type of Shape
W 24 x 76 W 14 x 26	W shape
S 24 x 100	S shape
M 8 x 18.5 M 10 x 9 M 8 x 34.3	M shape
C 12 x 20.7	American Standard Channel
MC 12 x 45 MC12 x 10.6	Miscellaneous Channel
HP 14 x 73	HP shape
L 6 x 6 x $\frac{3}{4}$	Equal Leg Angle
L 6 x 4 x $\frac{5}{8}$	Unequal Leg Angle
WT 12 x 38 WT 7 x 13	Structural Tee cut from W shape
ST 12 x 50	Structural Tee cut from S shape
MT 4 x 9.25 MT 5 x 4.5 MT 4 x 17.15	Structural Tee cut from M shape
PL $\frac{1}{2}$ x 18	Plate
Bar 1 ϕ	Square Bar
Bar 1$\frac{1}{4}$ ϕ	Round Bar
Bar 2$\frac{1}{2}$ x $\frac{1}{2}$	Flat Bar
Pipe 4 Std. Pipe 4 x-Strong Pipe 4 xx-Strong	Pipe
TS 4 x 4 x .375	Structural Tubing: Square
TS 5 x 3 x .375	Structural Tubing: Rectangular
TS 3 OD x .250	Structural Tubing: Circular

(Courtesy of American Institute of Steel Construction, Inc.)

MATERIAL KEY FOR PANEL AND REINFORCING TICKET

Item	Description	Quantity	How to Locate	Remarks
Back-up mat'l. (joints)	Size and material	Lin ft	₵ JT	
Blockout	Size and depth	—	₵ or edge	
Caulking	Type	Gallons	₵ JT	
Concrete	Mix #	Cu ft	—	
Drip, finish stop	Size	—	See remarks	
Fixtures	Type, supplier	No Pcs	₵	
Gaskets	Type, material supplier	Lin ft	Blow-up detail	
Glass	Size, type, supplier	Sq ft	'do'	
Hardware	Mark (refer to hardware detail), supplier (P/C or others)	No Pcs	₵	
Hole or opening	Size	—	₵ or edge	
Insert (coil or threaded)	Designation, supplier (P/C or others)	No Pcs	₵	
Insert (wedge type)	Designation, supplier (P/C or others)	No Pcs	See remarks	
Insulation	Material, thickness, size	Sq ft	Out to out	
Lifting loop	Diameter, length, material, projection	No Pcs	₵	
Rebar	Size, length, finish, grade	Lin ft	₵ or edge	
Rebar (bent)	Mark (make reference to bar list)	No Pcs	₵ or edge	
Reglet	Type, material, manufacturer	Lin ft	₵	
WWF	Size, finish, length x width, grade	Sq ft	₵	
WWF (bent)	Mark (refer to bar list)	Sq ft	₵	

No. of Courses C.B. Tile	S.F.U.	Brick	Height of Unit Plus Joint
		1	0-2$\frac{5}{8}$
	1	2	0-5$\frac{3}{8}$
1		3	0-8
	2	4	0-10$\frac{5}{8}$
		5	1-1$\frac{3}{8}$
2	3	6	1-4
		7	1-6$\frac{5}{8}$
	4	8	1-9$\frac{3}{8}$
3		9	2-0
	5	10	2-2$\frac{5}{8}$
		11	2-5$\frac{3}{8}$
4	6	12	2-8
		13	2-10$\frac{5}{8}$
	7	14	3-1$\frac{3}{8}$
5		15	3-4
	8	16	3-6$\frac{5}{8}$
		17	3-9$\frac{3}{8}$
6	9	18	4-0
		19	4-2$\frac{5}{8}$
	10	20	4-5$\frac{3}{8}$
7		21	4-8
	11	22	4-10$\frac{5}{8}$
		23	5-1$\frac{3}{8}$
8	12	24	5-4
		25	5-6$\frac{5}{8}$
	13	26	5-9$\frac{3}{8}$
9		27	6-0
	14	28	6-2$\frac{5}{8}$
		29	6-5$\frac{3}{8}$
10	15	30	6-8
		31	6-10$\frac{5}{8}$
	16	32	7-1$\frac{3}{8}$
11		33	7-4
	17	34	7-6$\frac{5}{8}$
		35	7-9$\frac{3}{8}$
12	18	36	8-0

No. of Courses C.B. Tile	S.F.U.	Brick	Height of Unit Plus Joint
		37	8-2$\frac{5}{8}$
	19	38	8-5$\frac{3}{8}$
13		39	8-8
	20	40	8-10$\frac{5}{8}$
		41	9-1$\frac{3}{8}$
14	21	42	9-4
		43	9-6$\frac{5}{8}$
	22	44	9-9$\frac{3}{8}$
15		45	10-0
	23	46	10-2$\frac{5}{8}$
		47	10-5$\frac{3}{8}$
16	24	48	10-8
		49	10-10$\frac{5}{8}$
	25	50	11-1$\frac{3}{8}$
17		51	11-4
	26	52	11-6$\frac{5}{8}$
		53	11-9$\frac{3}{8}$
18	27	54	12-0
		55	12-2$\frac{5}{8}$
	28	56	12-5$\frac{3}{8}$
19		57	12-8
	29	58	12-10$\frac{5}{8}$
		59	13-1$\frac{3}{8}$
20	30	60	13-4
		61	13-6$\frac{5}{8}$
	31	62	13-9$\frac{3}{8}$
21		63	14-0
	32	64	14-2$\frac{5}{8}$
		65	14-5$\frac{3}{8}$
22	33	66	14-8
		67	14-10$\frac{5}{8}$
	34	68	15-1$\frac{3}{8}$
23		69	15-4
	35	70	15-6$\frac{5}{8}$
		71	15-9$\frac{3}{8}$
24	36	72	16-0

No. of Courses C.B. Tile	S.F.U.	Brick	Height of Unit Plus Joint
		73	16-2$\frac{5}{8}$
	37	74	16-5$\frac{3}{8}$
25		75	16-8
	38	76	16-10$\frac{5}{8}$
		77	17-1$\frac{3}{8}$
26	39	78	17-4
		79	17-6$\frac{5}{8}$
	40	80	17-9$\frac{3}{8}$
27		81	18-0
	41	82	18-2$\frac{5}{8}$
		83	18-5$\frac{3}{8}$
28	42	84	18-8
		85	18-10$\frac{5}{8}$
	43	86	19-1$\frac{3}{8}$
29		87	19-4
	44	88	19-6$\frac{5}{8}$
		89	19-9$\frac{3}{8}$
30	45	90	20-0
		91	20-2$\frac{5}{8}$
	46	92	20-5$\frac{3}{8}$
31		93	20-8
	47	94	20-10$\frac{5}{8}$
		95	21-1$\frac{3}{8}$
32	48	96	21-4
		97	21-6$\frac{5}{8}$
	49	98	21-9$\frac{3}{8}$
33		99	22-0
	50	100	22-2$\frac{5}{8}$
		101	22-5$\frac{3}{8}$
34	51	102	22-8
		103	22-10$\frac{5}{8}$
	52	104	23-1$\frac{3}{8}$
35		105	23-4
	53	106	23-6$\frac{5}{8}$
		107	23-9$\frac{3}{8}$
36	54	108	24-0

No. of Courses C.B. Tile	S.F.U.	Brick	Height of Unit Plus Joint
		109	24-2$\frac{5}{8}$
	55	110	24-5$\frac{3}{8}$
37		111	24-8
	56	112	24-10$\frac{5}{8}$
		113	25-1$\frac{3}{8}$
38	57	114	25-4
		115	25-6$\frac{5}{8}$
	58	116	25-9$\frac{3}{8}$
39		117	26-0
	59	118	26-2$\frac{5}{8}$
		119	26-5$\frac{3}{8}$
40	60	120	26-8
		121	26-10$\frac{5}{8}$
	61	122	27-1$\frac{3}{8}$
41		123	27-4
	62	124	27-6$\frac{5}{8}$
		125	27-9$\frac{3}{8}$
42	63	126	28-0
		127	28-2$\frac{5}{8}$
	64	128	28-5$\frac{3}{8}$
43		129	28-8
	65	130	28-10$\frac{5}{8}$
		131	29-1$\frac{3}{8}$
44	66	132	29-4
		133	29-6$\frac{5}{8}$
	67	134	29-9$\frac{3}{8}$
45		135	30-0
	68	136	30-2$\frac{5}{8}$
		137	30-5$\frac{3}{8}$
46	69	138	30-8
		139	30-10$\frac{5}{8}$
	70	140	31-1$\frac{3}{8}$
47		141	31-4
	71	142	31-6$\frac{5}{8}$
		143	31-9$\frac{3}{8}$
48	72	144	32-0

C.B. = concrete block   S.F.U. = struct. face unit

DECIMAL EQUIVALENTS

Table X. DECIMALS OF AN INCH FOR EACH 64th OF AN INCH
With Millimeter Equivalents

Fractions	$\frac{1}{64}$ ths	Decimal	Millimeters (Approx.)	Fraction	$\frac{1}{64}$ ths	Decimal	Millimeters (Approx.)
...	1	.015625	0.397	...	33	.515625	13.097
$\frac{1}{32}$	2	.03125	0.794	$\frac{17}{32}$	34	.53125	13.494
...	3	.046875	1.191	...	35	.546875	13.891
$\frac{1}{16}$	4	.0625	1.588	$\frac{9}{16}$	36	.5625	14.288
...	5	.078125	1.984	...	37	.578125	14.684
$\frac{3}{32}$	6	.09375	2.381	$\frac{19}{32}$	38	.59375	15.081
...	7	.109375	2.778	...	39	.609375	15.478
$\frac{1}{8}$	8	.125	3.175	$\frac{5}{8}$	40	.625	15.875
...	9	.140625	3.572	...	41	.640625	16.272
$\frac{5}{32}$	10	.15625	3.969	$\frac{21}{32}$	42	.65625	16.669
...	11	.171875	4.366	...	43	.671875	17.066
$\frac{3}{16}$	12	.1875	4.763	$\frac{11}{16}$	44	.6875	17.463
...	13	.203125	5.159	...	45	.703125	17.859
$\frac{7}{32}$	14	.21875	5.556	$\frac{23}{32}$	46	.71875	18.256
...	15	.234375	5.953	...	47	.734375	18.653
$\frac{1}{4}$	16	.250	6.350	$\frac{3}{4}$	48	.750	19.050
...	17	.265625	6.747	...	49	.765625	19.447
$\frac{9}{32}$	18	.28125	7.144	$\frac{25}{32}$	50	.78125	19.844
...	19	.296875	7.541	...	51	.796875	20.241
$\frac{5}{16}$	20	.3125	7.938	$\frac{13}{16}$	52	.8125	20.638
...	21	.328125	8.334	...	53	.828125	21.034
$\frac{11}{32}$	22	.34375	8.731	$\frac{27}{32}$	54	.84375	21.431
...	23	.359375	9.128	...	55	.859375	21.828
$\frac{3}{8}$	24	.375	9.525	$\frac{7}{8}$	56	.875	22.225
...	25	.390625	9.922	...	57	.890625	22.622
$\frac{13}{32}$	26	.40625	10.319	$\frac{29}{32}$	58	.90625	23.019
...	27	.421875	10.716	...	59	.921875	23.416
$\frac{7}{16}$	28	.4375	11.113	$\frac{15}{16}$	60	.9375	23.813
...	29	.453125	11.509	...	61	.953125	24.209
$\frac{15}{32}$	30	.46875	11.906	$\frac{31}{32}$	62	.96875	24.606
...	31	.484375	12.303	...	63	.984375	25.003
$\frac{1}{2}$	32	.500	12.700	1	64	1.000	25.400

DECIMAL EQUIVALENTS OF 1'-0" FOR EACH $\frac{1}{8}$"

Inch	0	1	2	3	4	5	6	7	8	9	10	11
0	0	0.0833	0.1667	0.2500	0.3333	0.4167	0.5000	0.5833	0.6667	0.7500	0.8333	0.9167
$\frac{1}{8}$	0.0104	0.0938	0.1771	0.2604	0.3438	0.4271	0.5104	0.5938	0.6771	0.7604	0.8438	0.9271
$\frac{1}{4}$	0.0208	0.1042	0.1875	0.2708	0.3542	0.4375	0.5208	0.6042	0.6875	0.7708	0.8542	0.9375
$\frac{3}{8}$	0.0313	0.1146	0.1979	0.2813	0.3646	0.4479	0.5313	0.6146	0.6979	0.7813	0.8646	0.9479
$\frac{1}{2}$	0.0417	0.1250	0.2083	0.2917	0.3750	0.4583	0.5417	0.6250	0.7083	0.7917	0.8750	0.9583
$\frac{5}{8}$	0.0521	0.1354	0.2188	0.3021	0.3854	0.4688	0.5521	0.6354	0.7188	0.8021	0.8854	0.9688
$\frac{3}{4}$	0.0625	0.1458	0.2292	0.3125	0.3958	0.4792	0.5625	0.6458	0.7292	0.8125	0.8958	0.9792
$\frac{7}{8}$	0.0729	0.1563	0.2396	0.3229	0.4063	0.4896	0.5729	0.6563	0.7396	0.8229	0.9063	0.9896

INDEX